Electrospun Polymer Nanofibers for Food and Health Applications

Electrospun Polymer Nanofibers for Food and Health Applications

Special Issue Editor
Marija Gizdavic-Nikolaidis

MDPI • Basel • Beijing • Wuhan • Barcelona • Belgrade

Special Issue Editor
Marija Gizdavic-Nikolaidis
University of Auckland
New Zealand

Editorial Office
MDPI
St. Alban-Anlage 66
4052 Basel, Switzerland

This is a reprint of articles from the Special Issue published online in the open access journal *Fibers* (ISSN 2079-6439) in 2019 (available at: https://www.mdpi.com/journal/fibers/special_issues/electrospun_polymer_nanofibers_food_health).

For citation purposes, cite each article independently as indicated on the article page online and as indicated below:

LastName, A.A.; LastName, B.B.; LastName, C.C. Article Title. *Journal Name* **Year**, *Article Number*, Page Range.

ISBN 978-3-03928-192-3 (Pbk)
ISBN 978-3-03928-193-0 (PDF)

© 2020 by the authors. Articles in this book are Open Access and distributed under the Creative Commons Attribution (CC BY) license, which allows users to download, copy and build upon published articles, as long as the author and publisher are properly credited, which ensures maximum dissemination and a wider impact of our publications.

The book as a whole is distributed by MDPI under the terms and conditions of the Creative Commons license CC BY-NC-ND.

Contents

About the Special Issue Editor vii

Preface to "Electrospun Polymer Nanofibers for Food and Health Applications" ix

Chi-Ming Chiu, Jukkrit Nootem, Thanapich Santiwat, Choladda Srisuwannaket, Kornkanya Pratumyot, Wei-Chao Lin, Withawat Mingvanish and Nakorn Niamnont
Enhanced Stability and Bioactivity of *Curcuma comosa* Roxb. Extract in Electrospun Gelatin Nanofibers
Reprinted from: *Fibers* **2019**, *7*, 76, doi:10.3390/fib7090076 1

Nafeesa Mohd Kanafi, Norizah Abdul Rahman, Nurul Husna Rosdi, Hasliza Bahruji and Hasmerya Maarof
Hydrogel Nanofibers from Carboxymethyl Sago Pulp and Its Controlled Release Studies as a Methylene Blue Drug Carrier
Reprinted from: *Fibers* **2019**, *7*, 56, doi:10.3390/fib7060056 13

Mirei Tsuge, Kanoko Takahashi, Rio Kurimoto, Ailifeire Fulati, Koichiro Uto, Akihiko Kikuchi and Mitsuhiro Ebara
Fabrication of Water Absorbing Nanofiber Meshes toward an Efficient Removal of Excess Water from Kidney Failure Patients
Reprinted from: *Fibers* **2019**, *7*, 39, doi:10.3390/fib7050039 29

Marah Trabelsi, Al Mamun, Michaela Klöcker, Lilia Sabantina, Christina Großerhode, Tomasz Blachowicz and Andrea Ehrmann
Increased Mechanical Properties of Carbon Nanofiber Mats for Possible Medical Applications
Reprinted from: *Fibers* **2019**, *7*, 98, doi:10.3390/fib7110098 38

Eri Niiyama, Kanta Tanabe, Koichiro Uto, Akihiko Kikuchi and Mitsuhiro Ebara
Shape-Memory Nanofiber Meshes with Programmable Cell Orientation
Reprinted from: *Fibers* **2019**, *7*, 20, doi:10.3390/fib7030020 48

About the Special Issue Editor

Marija Gizdavic-Nikolaidis graduated with a BSc (Honours) in physical chemistry at the Faculty of Physical Chemistry, University of Belgrade in 1995. She obtained her PhD in chemistry at the University of Auckland in 2005. She has 20+ years research experience in materials and polymer science fields working with both academic and industrial groups worldwide. Dr. Gizdavic-Nikolaidis is a Senior Research Fellow at the University of Auckland and elected Fellow of New Zealand Institute of Chemistry. She has been actively engaged as a research innovator and science leader in cross-disciplinary polymer chemistry research at the University of Auckland, focusing on linking academic research with New Zealand industry needs.

Preface to "Electrospun Polymer Nanofibers for Food and Health Applications"

This Special Issue, titled "Electrospun Polymer Nanofibers for Food and Health Applications" (https: // www.mdpi.com/ journal/ fibers/ special_issues/ electrospun_polymer_nanofibers_food_ health), covers current research in this field, with a particular focus on the development of electrospun nanofiber meshes for various medical applications. Electrospun PCL-based polyurethane nanofiber meshes exhibit advanced shape-memory abilities and good biocompatibility. These shape-memory nanofier meshes can with further study offer insight into the effect of dynamic matrix structure changes on cell behaviors. Furthermore, an additional benefit of the electrospinning method is the opportunity for producing nanofibers using natural or synthetic polymers, which can be used in drug and therapeutic agent deliveries, wound dressings, and in tissue engineering. Electrospun nanofibers of carboxymethyl sago pulp (CMSP) extracted from sago waste in combination with poly(ethylene oxide) (PEO) can be utilized as a methylene blue drug carrier. ZnO-blended polyacrylonitrile (PAN) nanofiber mats have remarkably increased fiber diameters, resulting in improved mechanical properties appropriate for tissue engineering applications. Electrospun gelatin nanofibrous mats can stabilize the bioactive compounds in Curcuma comosa Roxb. extract (CE) under an accelerated storage condition. Moreover, these electrospun nanofibers showed the capacity of the release of bioactive ingredients having antioxidant properties and, thus, it could potentially be used for face mask application. In addition, poly(sodium acrylate), PSA electrospun nanofiber mesh has the potential to be used to remove excess water from the bloodstream without requiring specialized equipment.

Marija Gizdavic-Nikolaidis
Special Issue Editor

Article

Enhanced Stability and Bioactivity of *Curcuma comosa* Roxb. Extract in Electrospun Gelatin Nanofibers

Chi-Ming Chiu [1,2], Jukkrit Nootem [1], Thanapich Santiwat [1], Choladda Srisuwannaket [1], Kornkanya Pratumyot [1], Wei-Chao Lin [2], Withawat Mingvanish [1] and Nakorn Niamnont [1,*]

[1] Organic Synthesis, Electrochemistry & Natural Product Research Unit, Department of Chemistry, Faculty of Science, King Mongkut's University of Technology Thonburi, Bangkok 10140, Thailand
[2] Department of Cosmetic Science, College of Pharmacy and Science, Chia Nan University of Pharmacy and Science, Tainan 71710, Taiwan
* Correspondence: nakorn.nia@kmutt.ac.th

Received: 3 July 2019; Accepted: 10 August 2019; Published: 27 August 2019

Abstract: Electrospun fiber can be used as a carrier for releasing active ingredients at the target site to achieve the effects of drug treatment. The objectives of this research work were to study suitable conditions for producing electrospun gelatin fiber loaded with crude *Curcuma comosa* Roxb. extract (CE) and to study antioxidant, anti-tyrosinase and anti-bacterial activities and its freeze–thaw stability as well. To achieve optimal conditions for producing electrospun gelatin fiber, the concentration of gelatin was adjusted to 30% w/v in a co-solvent system of acetic acid/water (9:1 v/v) with a feed rate of 3 mL/h and an applied voltage of 15 kV. The lowest percent loading of 5% (w/v) CE in gelatin nanofiber exhibited the highest DPPH radical scavenging activity of 94% and the highest inhibition of tyrosinase enzyme of 35%. Moreover, the inhibition zones for antibacterial activities against *S. aureus* and *S. epidermidis* were 7.77 ± 0.21 and 7.73 ± 0.12 mm, respectively. The freeze–thaw stability of CE in electrospun gelatin nanofiber was significantly different ($p < 0.05$) after the 4th cycle as compared to CE. Electrospun gelatin nanofiber containing CE also showed the capacity of the release of bioactive ingredients possessing anti-oxidant properties and, therefore, it could potentially be used for face masks.

Keywords: *Curcuma comosa*; electrospinning; gelatin; *S. epidermidis*; *S. aureus*

1. Introduction

Curcuma comosa Roxb. is a species in the genus *Curcuma* of the family Zingiberaceae. It has been used in traditional folk medicine for the treatment of a wide range of diseases in many Asian countries [1,2]. The individual compounds isolated from *C. comosa* have displayed various biological properties such as estrogenic [3], choleretic [4], nematocidal [5], anti-inflammatory [6], antioxidant [7], and antibacterial activities [8]. The active constituents of *C. comosa* extract have been identified as acetophenones, curcuminoids, sesquiterpenes, and diarylheptanoids [9]. Additionally, the major bioactive compounds were diarylheptanoid derivatives at a level of 80–90% in crude extract [10]. Amongst non-phenolic diarylheptanoids, (3R)-1,7-diphenyl-(4E,6E)-4,6-heptadien-3-ol exhibited the highest anti-inflammatory [11], anti-oxidant [12,13] antibacterial activities [14] and anti-tyrosinase. However, its backbone consisting of double bonds is easily oxidized and transformed upon exposure to oxygen, light, and temperature, which decreases its bioactivities [15].

Various encapsulation techniques have been employed to improve the stability of diarylheptanoids using lipid nanoemulsions, beta-cyclodextrin, and calcium alginate [16–18]. These have demonstrated an increase in the stability and bioactivity of diarylheptanoids under various environmental conditions.

However, encapsulation is associated with several limitations such as handling, drug-loading, and upscaling. Therefore, an alternative encapsulation method, known as electrospinning, has been developed to produce continuous fibers in levels of submicron diameters down to nanometer diameters by transforming polymer solutions or polymer melts to nonwoven-fibrous materials using electricity. Furthermore, an electrospun nanofibrous mat exhibited superior properties such as effective drug encapsulation, high drug loading efficacy, wide range of polymers to accommodate compatibility, and the ability to adjust drug releasing level through its high surface to volume ratio [19–21].

Several polymeric materials such as cellulose [22–24], polycaprolactone [25], poly (lactic acid) [26], polyvinyl alcohol [27,28] and gelatin [29,30] have been currently used to encapsulate many drugs or bioactive compounds by electrospinning. Gelatin, a hydrophilic polymer, has commonly been electrospun together with other polymers to carry bioactive compounds due to its biocompatible, nontoxic and biodegradable characteristics [31]. Many research studies have confirmed that bioactive compounds loaded into electrospun gelatin fibers, including silver nanoparticles [32], antibiotic drugs [33], and vitamins A and E [34] showed significant releasing property, increased bioactivity, and highly enhanced stability. Herein, we reported for the first time on the optimal conditions for the fabrication of the electrospun fibers of gelatin loaded with crude *Curcuma comosa* Roxb extract (CE). The bioactivities of the electrospun fibers containing bioactive compounds were also investigated using 2,2-diphenyl-1-picrylhydrazyl (DPPH) radical scavenging activity assay, anti-tyrosinase activity assay and antibacterial activity assay. Moreover, its freeze–thaw stability testing was carried out.

2. Experiments

2.1. Reagents and Apparatus

Gelatin powder from porcine skin with the gel strength of ~300 g Bloom (Type A), 2,2-diphenyl-1-picrylhydrazyl 3,4-dihydroxy-L-phenylalanine (DPPH; ≥98%), phosphate-buffered saline (PBS), 3-(3,4-Dihydroxyphenyl)-L-alanine (L-DOPA), and tyrosinase were purchased from Sigma–Aldrich (St. Louis, MO, USA). All analytical grade reagents of methanol, ethanol, acetone, acetic acid, and ethyl acetate (EA) with 99% purification were obtained from RCI Labscan (Thailand). The other chemicals used were of analytical grade. Powder of *Curcuma comosa* Roxb. rhizome was purchased from a local Thai Herb Trading market in Bangkok, Thailand in May 2016. Muller-Hinton Agar (MHA), Gentamicin, dimethyl sulfoxide (DMSO), *Staphylococcus aureus*, and *Staphylococcus epidermidis* were all supplied by the laboratory at Rangsit University, in Pathum Thani, Thailand.

2.2. Preparation of Crude Ethanol Extract of Curcuma comosa Roxb. Rhizome Powder

A total of 500 g of powder of *Curcuma comosa* Roxb. rhizome was macerated with 2 × 500 mL of 95% ethanol at room temperature for 3 days. The ethanol extract solution was then concentrated by using a rotary evaporator to obtain the crude ethanol extract of *Curcuma comosa* Roxb. rhizome (CE). CE was kept at 4 °C with light protection. The amounts of diarylheptanoid derivatives in CE were measured using high performance liquid chromatography (HPLC, Agilent Technologies, CA, USA) at detector wavelength of 259 nm [35]. The purity of diarylheptanoid derivatives was found to be 89.3% by HPLC technique.

2.3. Identification of Bioactive Compounds of the Crude Ethanol Extract of Curcuma comosa Roxb. Rhizome Powder

Preliminary phytochemical testing was used to detect the presence of plant secondary metabolites, i.e., alkaloids, flavonoids, saponins, steroids, tannins, phenols according to Prashanth et al.'s method [36] with little modifications. Furthermore, the crude ethanol extract of *Curcuma comosa* Roxb. rhizome (CE) powder was characterized by Fourier-transform infrared spectrometer (FTIR, Thermo scientific, CA, USA) in the region of 4000 to 600 cm^{-1} and 400 MHz ^{1}H, and 100 MHz ^{13}C-NMR spectrometer (Avance III HD 400 MHz of NMR, Bruker, Zurich, Switzerland) for determining the presence of diarylheptanoid derivatives in CE. The deuterated solvent for NMR analysis was chloroform-d ($CDCl_3$).

2.4. Electrospinning and Characterization of Gelatin Nanofibrous Mat

The gelatin solutions of known concentrations (20, 25 and 30% (w/v)) were prepared in a mixed solvent system containing acetic acid and H_2O (9:1 (v/v)). Each concentration of the gelatin solution was then loaded into a 5-mL syringe connected with a metal needle of inner diameter of 0.22 mm. The needle was connected to the positive electrode (Gamma high voltage research FL 32174, United States) and the negative electrode was connected to the rotary spinning collector. The distance of needle tip away from the collector was located about 15 cm. Each gelatin solution was injected under the applied voltage of 15 kV at a flow rate of 3 mL/h and nanofibers were collected on the rotary spinning collector at 1200 rpm [24]. In order to study morphology of electrospun nanofibers fabricated under various experimental electrospinning conditions, all nanofiber samples were dried in a vacuum oven at 50 °C for 3 h. After that, each sample was cut into a small piece with a size of 1×1 cm, which was coated with gold before examined using a scanning electron microscope (JSM-6610 LV; SEM, JEOL, Tokyo, Japan). The diameters of each electrospun fiber were then determined using ImageJ 1.51 K. software ($n \geq 100$) [29,30].

2.5. Studies of Characteristics of Electrospun Gelatin Nanofibers Loaded with the Crude Ethanol Extract of Curcuma comosa Roxb. Rhizome

Percent loading. To determine the suitable content of the crude ethanol extract of *Curcuma comosa* Roxb. rhizome (CE) loaded into electrospun gelatin nanofibrous mat, a predetermined weight of CE (1, 5, 10, 15% (w/v)) was dissolved in 2 g of gelatin solution (30% and the solution mixture was electrospun under optimum conditions. Crude diarylheptanoid derivatives extract (1, 5, 10, 15% (w/v)) of a predetermined weight was dissolved in 2 g gelatin solution to generate electrospun fiber under optimum conditions. 10 mg of a gelatin nanofibrous mat loaded with CE was then re-dissolved in 2 mL of methanol. The content of diarylheptanoids in the solution mixture was determined using a UV-Vis spectrometer at 254 nm and 285 nm based on the method of Yingngam et al. [10] and reported as mean value $\pm$ SD ($n = 3$).

Anti-oxidant activity. 50 mg of a electrospun gelatin nanofibrous mat containing various concentrations of the crude ethanol extract of *Curcuma comosa* Roxb. rhizome (CE) (1, 5, 10 and 15% w/v) was re-dissolved in 5 mL of methanol by aid of sonication machine for 5 min and was subjected to examine its DPPH radical scavenging activity in accordance to the method of Devi et al. [37] with some modifications. 100 µL of each sample was mixed with 100 µL of 0.5 mM DPPH solution in methanol. The solution mixture was incubated in the dark at room temperature for 15 min. The absorbance of the solution mixture was measured at 517 nm using Perkin-Elmer multimode plate reader. All tests were done in triplicate.

The percentage of DPPH radical scavenging activity was calculated as follows:

$$\text{DPPH scavenging activity (\%)} = \left[\frac{A_0 - (A_1 - A_2)}{A_0} \right] \times 100 \qquad (1)$$

where A_0 was the absorbance of the blank solution; A_1 was the absorbance of the nanofiber solution with DPPH; A_2 was the absorbance of the nanofiber solution without DPPH.

Anti-tyrosinase activity. The anti-tyrosinase activity of electrospun gelatin nanofibrous mat loaded with various concentrations of the crude ethanol extract of *Curcuma comosa* Roxb. rhizome (CE) was carried out based on inhibition of mushroom tyrosinase-catalyzed oxidation of L-DOPA. According to Masamoto et al.'s [38] method with some modifications, 30 µg of each electrospun gelatin nanofibrous mat loaded with various concentrations of CE was dissolved in 3 mL of sodium phosphate buffer pH 6.8 and 1.0 mL of a 1.5 mM L-DOPA, 0.1 mL of DMSO with or without a test sample and 1.8 mL of a phosphoric acid (pH 6.8) were mixed well in a test tube. The reaction mixture was preincubated at 25 °C for 10 min, before 0.1 mL of an aqueous tyrosinase solution (100 U/mL, Sigma Chemical Co. (St. Louis, MO, USA)) was added into the reaction mixture and the reaction was monitored at 475 nm. The control reaction was conducted with DMSO. Each measurement was performed at least in triplicate.

The inhibitory percentage of tyrosinase activity was calculated using the following formula:

$$\text{Tyrosinase inhibitory activity } (\%) = \left[\frac{(A_1 - A_2) - (A_3 - A_4)}{(A_1 - A_2)}\right] \times 100 \qquad (2)$$

where A_1 was the absorbance of the blank solution with enzyme; A_2 was the absorbance of the blank solution without enzyme; A_3 was the absorbance of the nanofiber solution with enzyme; A_4 was the absorbance of the nanofiber solution without enzyme.

Anti-bacterial activity. The initial evaluation of anti-bacterial activity of electrospun nanofibrous mat loaded with various concentrations of the crude ethanol extract of *Curcuma comosa* Roxb. rhizome (CE) was carried out using agar diffusion method (Kirby-Bauer test). The anti-bacterial activity testing was done by following the Clinical and Laboratory Standard Institute (CLSI) Guidelines 2017 with some modifications. Two Gram-positive bacteria; *Staphylococcus aureus* and *Staphylococcus epidermidis*, were used for this study. Briefly, electrospun gelatin nanofiber sheets loaded and unloaded 500–1000 µg/mL CE and 0.25 µg/mL gentamicin (positive control) were dissolved in DMSO. Paper discs (6 mm in diameter) were soaked with 5 µL of test samples for 15 min and let them dry in a laminar air flow cabinet for 2 h. After that, they were placed on Mueller-Hinton agar (MHA) plates adjusted to pH 7.2–7.4, which have been inoculated with test bacteria at 37 °C for 18 h. The diameters of inhibition zones (mm) were measured after 18 h. Filter papers containing only DMSO were used a control and no inhibition zones were not observed. Testing was performed in triplicate and repeated three times. The minimum inhibitory concentration (MIC) of electrospun nanofibrous mat loaded with various concentrations of CE was carried out using agar dilution method, which also followed the Clinical & Laboratory Standard Institute (CLSI) Guidelines 2017 with some modifications. The MIC value of the test sample was determined as the lowest concentration of the test sample that completely prevent any turbidity or growth of the test bacteria. All test samples were performed in triplicate.

2.6. Stability of Electrospun Gelatin Nanofibrous Mat Loaded with Crude Ethanol Extract of Curcuma comosa Roxb. Rhizome Powder

The stability of electrospun gelatin nanofibrous mat loaded by crude ethanol extract of *Curcuma comosa* Roxb. rhizome powder (CE) have been evaluated by using freeze–thaw cycle test via their remaining content of bioactive compounds in CE and remaining DPPH radical scavenging activity [34]. 30 mg of nanofibrous mat mixed with CE dissolved in 3 mL of methanol and an only CE solution in methanol at the same concentration were prepared. All test solutions from gelatin-CE nanofibrous mats and CE solution were freezed in freezer at −5 °C for 16 h and then thawed in a conventional oven at 45 °C for 8 h. After that, all test solutions were determined for the remaining contents of bioactive compounds in CE by comparison of calibration curves of CE using an UV-Vis spectrophotometer at 254 and 285 nm. The remaining DPPH scavenging activity of each test sample at any freeze–thaw cycle were assessed as mentioned before. The freeze/thaw cycle testing were repeated up to 20 times (a duration of around 3 weeks).

2.7. Statistical Analysis

All experiments were carried out in triplicate. The experimental results are reported as mean ± standard deviation (SD). One-way analysis of variance (ANOVA) was performed for the triplicate ($n = 3$) testing data using SigmaPlot 10.0.

3. Results and Discussion

3.1. Extraction and Characterisation of Diarylheptanoid Derivatives in Crude Ethanol Extract of Curcuma comosa Roxb. Rhizome Powder

Crude ethanol extract of *Curcuma comosa* Roxb. rhizome powder (CE) was obtained as viscous and dark brown solid with 5.37% yield (w/w) of dried plant material. The presence of diarylheptanoid derivatives (Figure 1) and bioactive compounds in CE was studied using spectroscopic techniques (Figures S1–S3 in Supplementary Data). Furthermore, the quantity of diarylheptanoid derivatives in CE was estimated to be 89.3% by using HPLC. The ^{1}H-NMR spectrum of CE (Figure S1) showed important peaks at 6.5–7.5 ppm (Ar–H; aromatic protons), 3.3–5.0 ppm (H–Csp^3–OH; protons attached to carbon atoms singly bonded to oxygen), and 2.0–3.0 ppm (HCsp^3-C; aliphatic protons of sp^3 hybridized carbon atoms of diarylheptanoids). In addition, the results of preliminary phytochemical screening testing presented in Table S1 in the Supplementary Data revealed the existence of other secondary metabolites in CE, which may include cardiac glycosides, phenols, flavonoids, tannins, and terpenoids. These results confirm the phytochemical evidences reported previously for CE [35,36].

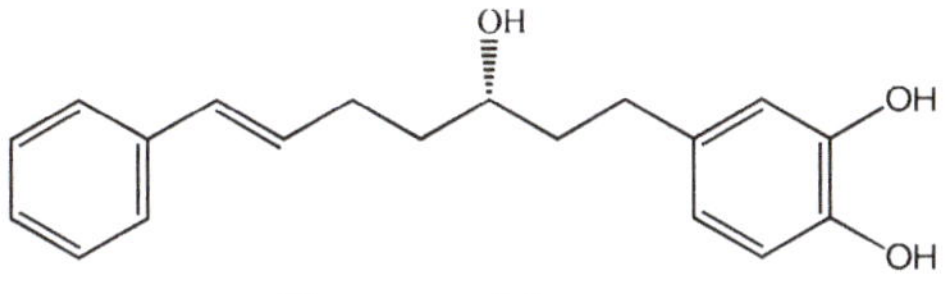

Compound 1

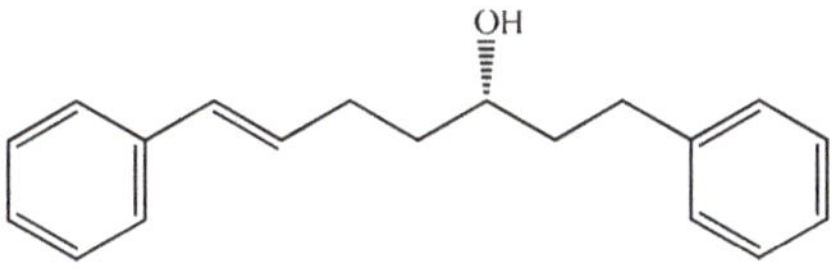

Compound 2

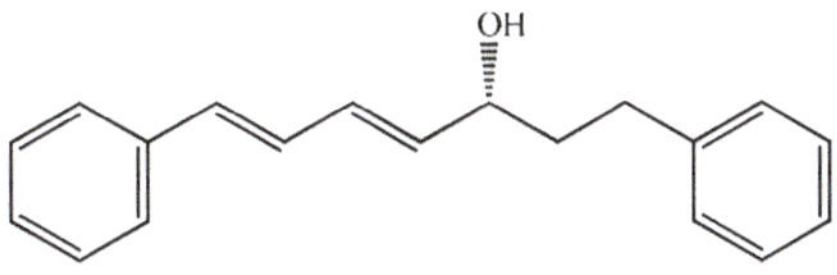

Compound 3

Figure 1. Chemical structures of diarylheptanoids (compound 1: (3S)-1-(3,4-Dihydroxyphenyl)-7-phenyl-(6E)-6-hepten-3-ol; compound 2: (3R)-1,7-Diphenyl-(4E,6E)- 4,6-heptadiene-3-ol; compound 3: (3S)-1,7-Diphenyl-(6E)-6-hepten-3-ol) in crude ethanol extract of *Curcuma comosa* Roxb. rhizome powder.

3.2. Effect of Gelatin Concentrations on the Morphology of Electrospun Gelatin Nanofibers

The acetic acid/water co-solvent system at the volumetric ratio of 9:1 was chosen for fabricating gelatin nanofibers by using electrospinning technique [29,30]. The morphology of electrospun gelatin nanofibers fabricated from gelatin solutions in acetic acid/water (9:1 v/v) at various gelatin concentrations was examined using scanning electron microscopy (SEM) From SEM image (Figure 2), the average diameters of electrospun nanofibers obtained from 20, 25, and 30% (w/v) gelatin solutions were 183 ± 29, 367 ± 64 and 547 ± 94 nm, respectively. This may be because polymer molecules and solvents affect the surface tension and the proper evaporation rate of the solution to produce smooth and uniform fibers. Although 20% (w/v) gelatin solution gave electrospun nanofibers the finest average diameter (mean = 183.4 nm), its nanofibers has relatively low toughness. From SEM micrographs of electrospun nanofibers, gelatin concentrations lower than 25% (w/v) could easily cause the cracking of nanofibers. However, cracking is less likely to be seen in electrospun nanofibers obtained from 30% (w/v) gelatin solution, and its average diameter is relatively in range of 500–600 nm. Therefore, a 30% (w/v) gelatin concentration in acetic acid/water (9:1 v/v) was selected as the optimal condition in this study.

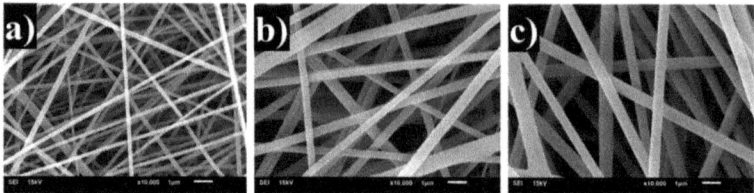

Figure 2. Electrospun gelatin nanofibers fabricated from gelatin solutions in acetic acid/water (9:1 (v/v)) at various gelatin concentrations: (**a**) 20%, (**b**) 25%, and (**c**) 30% (w/v).

3.3. Effect of Various Concentrations of Crude Ethanol Extract of Curcuma comosa Roxb. Rhizome Powderloaded on the Morphology of Electrospun Gelatin Nanofibers

The optimal conditions for electrospinning were chosen to fabricate gelatin nanofibrous mat containing crude ethanol extract of *Curcuma comosa* Roxb. rhizome powder (CE) with various concentrations (1, 5, 10 and 15% (w/v)). Figure 3 showed that the gelatin nanofibers loaded with 5% (w/v) of CE were relatively with an average diameter of 1089 ± 611.38 nm. However, 10 and 15% (w/v) of CE made gelatin solutions more viscous and resulted in an increase in the average diameters of gelatin nanofibers, which might cause to the effect of the high CE extract concentration with the increase viscosity, and reduced conductivity to main affect in an increase the average diameters of the electrospun gelatin nanofibrous mat. Thus, the addition of 5% (w/v) of CE into 30% w/v of gelatin solution is the most suitable condition in this study.

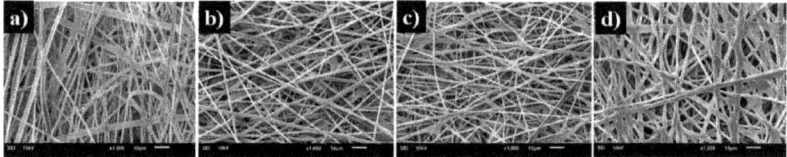

Figure 3. Electrospun gelatin nanofibers with various concentrations of crude ethanol extract of *Curcuma comosa* Roxb.rhizome powder: (**a**) 1%, (**b**) 5%, (**c**) 10%, and (**d**) 15% (w/v) in acetic acid/H_2O (9:1 (v/v)).

3.4. Effect of Percent Loading on Anti-Oxidant Activity of Electrospun Gelatin Nanofibrous Mat Loaded with Crude Ethanol Extract of Curcuma comosa Roxb. Rhizome Powder

The effect of percent loading of crude ethanol extract of *Curcuma comosa* Roxb. rhizome powder (CE) on anti-oxidant activity of electrospun gelatin nanofibers, which were fabricated from 30% (w/v) gelatin solution in acetic acid/H_2O (9:1 (v/v)) was explored by DPPH radical scavenging assay. Figure 4 presents the plot of percent loading of CE against DPPH radical scavenging rate. The percent loadings at 1, 5, 10, and 15% w/v yielded DPPH radical scavenging rates of 35%, 94%, 94%, and 94.5%, respectively. However, the DPPH radical scavenging rates at the percent loadings of 5, 10 and 15% w/w, they were not statistically significant ($p < 0.05$). Therefore, to achieve significant free radical scavenging activity, the lowest percent loading with the highest DPPH radical scavenging rate of 5% w/v was preferred. This showed that CE was a good source of natural antioxidants.

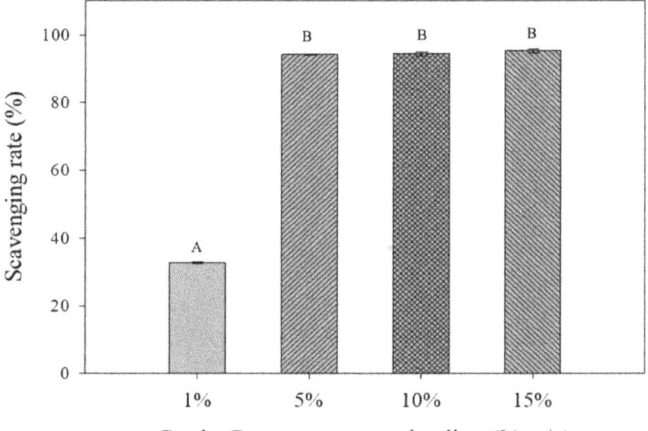

Figure 4. Effect of percent loading of crude ethanol extract of *Curcuma comosa* Roxb. powder in electrospun gelatin nanofibers on DPPH radical-scavenging activity. [A,B] Means with different capital letter superscripts are significantly different ($p < 0.05$).

3.5. Effect of Percent Loading on Anti-Tyrosinase Activity of Electrospun Gelatin Nanofibrous Mat Loaded with Crude Ethanol Extract of Curcuma comosa Roxb. Rhizome Powder

Figure 5 is the plot of anti-tyrosinase activity of electrospun gelatin nanofibrous mat against percent loading of crude ethanol extract of *Curcuma comosa* Roxb. rhizome powder (CE). The results showed that, at the percent loading of 1% (w/v) of CE, the inhibition rate of tyrosinase enzyme was 30%. As the percent loading of CE increased from 5% to 15% (w/v), the inhibition rate of tyrosinase enzyme slightly increased from 35% to 37%. The lowest percent CE loading with the highest anti-tyrosinase activity of 5% w/v was therefore selected due to its good performance in inhibiting melanin production.

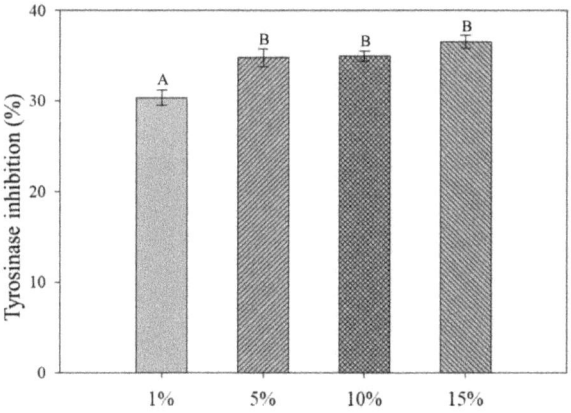

Figure 5. Effect of percent loading of crude ethanol extract of *Curcuma comosa* Roxb. powder in electrospun gelatin nanofibers on anti-tyrosinase activity. [A,B] Means with different capital letter superscripts are significantly different ($p < 0.05$).

3.6. Effect of Percent Loading on Antibacterial Activity of Electrospun Gelatin Nanofibrous Mat Loaded with Crude Ethanol Extract of Curcuma comosa Roxb. Rhizome Powder

Electrospun gelatin nanofiber sheets loaded with various concentrations of crude ethanol extract of *Curcuma comosa* Roxb. rhizome powder (CE) were tested for their antibacterial activity by the disc diffusion method and by the broth dilution method. Gentamicin was used for this study as a positive control. It gave inhibitory activity against *S. aureus* with the inhibition zone of 24.30 ± 0.00 mm and MIC of 0.25 µg/mL and against *S. epidermidis* with the inhibition zone of 30.50 ± 0.17 mm and MIC of 0.25 µg/mL. The results of the antibacterial studies with regard to the zone of inhibition are shown in Table 1. The electrospun gelatin nanofibrous mat without CE loading (blank nanofiber) showed no inhibitory activity against *S. aureus* and *S. epidermidis*. CE solution showed inhibitory activity against *S. aureus* with the inhibition zone of 9.80 ± 0.17 mm and MIC of 1000 µg/mL and against *S. epidermidis* with the inhibition zone of 9.93 ± 0.06 mm and MIC of 500.0 µg/mL (Figure 6). The electrospun gelatin nanofibrous mat with CE loading of 5% (w/v) (CE nanofiber) showed inhibitory activity against *S. aureus* with the inhibition zone of 7.77 ± 0.21 mm and against *S. epidermidis* with the inhibition zone of 7.73 ± 0.12 mm (Table 1). This means that CE nanofiber could inhibit *S. epidermidis* and *S. aureus* in median level of inhibition. This may be due to the good adhesion of the bioactive compounds in electrospun gelatin nanofibrous mat. Although, the limit of maximum loading of CE concentration in nanofibrous mat of the work was low concentration level of bioactive compounds. Additionally, CE nanofiber could be the efficacy of the antimicrobial treatment for atopic dermatitis (AD, eczema) as a common inflammatory skin.

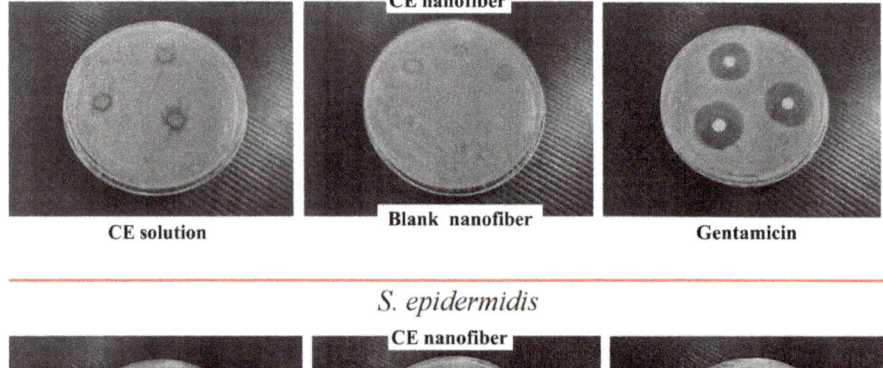

Figure 6. Effect of 5% w/v crude extract of electrospun fiber on antibacterial activity inhibition. Antibacterial activity of crude ethanol extract of *Curcuma comosa* Roxb. rhizome powder (CE) in electrospun gelatin nanofibrous mat; 5% (w/v) CE solution, blank nanofiber, 5% (w/v) CE nanofiber, and gentamicin positive control (5% (w/v)). Test disc diameter is 6 mm.

Table 1. Inhibition zone and MIC value of 5% (w/v) crude ethanol extract of *Curcuma comosa* Roxb. rhizome powder crude extract on electrospun gelatin nanofiber sheets.

Sample	S. aureus		S. epidermidis	
	Inhibition Zone [a] (mm)	MIC (μg/mL)	Inhibition Zone [a] (mm)	MIC (μg/mL)
CE solution	9.80 ± 0.17	1000	9.93 ± 0.06	500
Blank nanofiber	NA *	-	NA *	-
CE nanofiber	7.77 ± 0.21	-	7.73 ± 0.12	-
Gentamicin	24.30 ± 0.00	0.25	30.50 ± 0.17	0.25

* NA = no activity, [a] include disc diameter (6.0 mm).

3.7. Stability of Crude Extract-Loaded Gelatin Nanofibrous Mat

The stability of crude ethanol extract of *Curcuma comosa* Roxb. rhizome powder (CE) loaded into electrospun gelatin nanofibrous mat was investigated using a freeze–thaw cycle test. The concentrations of bioactive compounds released from electrospun gelatin nanofibrous mats were determined by following the method's Li [33]. At the 20th cycle, the released concentration of bioactive compounds from tested electrospun gelatin nanofibrous mat was 620 mg/L, which was significantly higher than the remaining concentration of CE in the tested solution (540 mg/L) (Figure 7). This confirmed that the bioactive compounds loaded into the gelatin nanofiber showed increasing stability. Moreover, Figure 8 showed that the DPPH radical scavenging activity of both free CE and CE loaded into electrospun gelatin nanofibrous mat decreased rapidly with progressing freeze–thaw cycles. However, free CE exhibited decreasing activity faster than CE loaded into electrospun gelatin nanofibrous mats. At the 20th cycle, the electrospun gelatin nanofibrous mats loaded with CE had the DPPH radical scavenging rate of 93% whereas the free CE had the DPPH radical scavenging rate of 89%. The results clearly indicate that electrospun gelatin nanofibrous mat can be used as a drug carrier for CE.

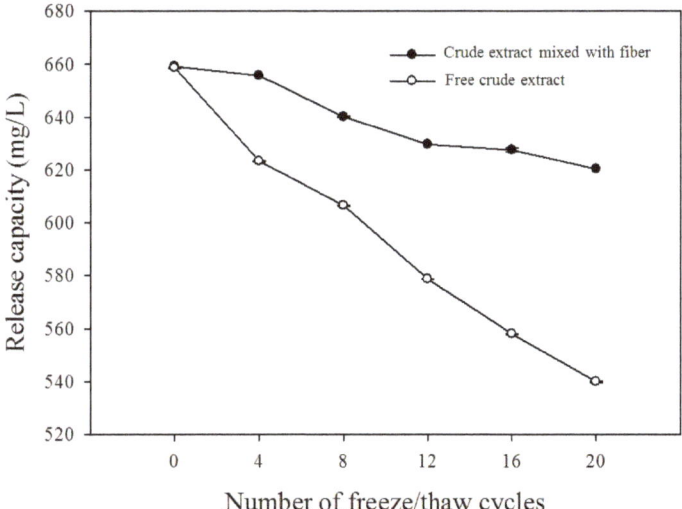

Figure 7. Change in concentrations of bioactive compounds in crude ethanol extract of *Curcuma comosa* Roxb. rhizome powder (CE) released from electrospun gelatin nanofibrous mats during freeze–thaw cycle testing; (●): electrospun gelatin nanofibrous mats loaded with 5% (w/v) of CE and (o): 5% (w/v) of only CE solution.

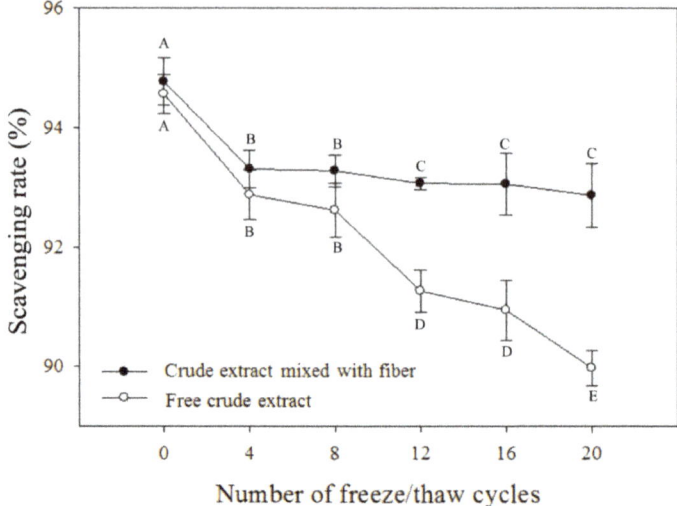

Figure 8. Change in percent of DPPH radical scavenging activity of bioactive compounds in crude ethanol extract of *Curcuma comosa* Roxb. rhizome powder (CE) released from electrospun gelatin nanofibrous mat during freeze–thaw cycle testing; (•): electrospun gelatin nanofiber sheets loaded with 5% (w/v) of CE and (o): 5% (w/v) of CE solution. ^{A,B,C,D,E} Means with different capital letter superscripts are significantly different ($p < 0.05$).

4. Conclusions

The suitable conditions for fabricating electrospun gelatin nanofibrous mat loaded with crude ethanol extract of *Curcuma comosa* Roxb. rhizome powder (CE) were 30% (w/v) gelatin solution in acetic acid/H_2O (9:1 (v/v)) mixed with 5% (w/v) of CE. The electrospun gelatin nanofibrous mat loaded with 5% (w/v) of CE showed significant antioxidant activity, and anti-tyrosinase activity. In addition, it exhibited antibacterial activity against *S. aureus* and *S. epidermidis* as compared to the positive control gentamicin. Notably, the electrospun gelatin nanofibrous mat was proved to stabilize the bioactive compunds in CE under an accelerated storage condition. The results showed that drugs loaded into electrospun gelatin nanofibers were appropriate for facial mask applications.

Supplementary Materials: The following are available online at http://www.mdpi.com/2079-6439/7/9/76/s1, Figure S1: 400 MHz ^{1}H-NMR spectrum of crude *Curcuma comosa* Roxb. extract., Table S1: Results of premilinary phytochemical screening of the crude ethanol extract of *Curcuma comosa* Roxb.

Author Contributions: Conceptualization, N.N. and W.M.; data curation, C.-M.C. and N.N.; formal analysis, C.-M.C. and W.M.; funding acquisition, N.N.; investigation, C.-M.C., J.N. and T.S.; methodology, N.N., C.-M.C., J.N. and T.S.; project administration, N.N.; supervision, N.N. and W.M.; writing—original draft, C.-M.C.; and writing—review and editing, N.N., W.M., C.S., W.-L.L. and K.P.

Funding: This research was funded by the Thailand research fund [MRG6180070], the research fund for DPST graduate with first placement, Thai government [No.11/2557] and the APC was funded by the KMUTT 55th Anniversary Commemorative Fund. C. C gratefully acknowledges the scholarship of Faculty of Science, King Mongkut's University of Technology Thonburi.

Acknowledgments: The authors would like to thanks Wei-Jing Hung and Wei-Chao Lin, Department of Cosmetic Science, Chia Nan University of Pharmacy and Science for support and encouragement.

Conflicts of Interest: The authors declare no conflicts of interest.

References

1. Saensouk, S.; Saensouk, P.; Pasorn Chantaranothai, P. Diversity and uses of zingiberaceae in nam nao national park, chaiyaphum and phetchabun provinces, Thailand, with a new record for Thailand. *Agric. Nat. Resour.* **2016**, *50*, 445–453. [CrossRef]
2. Su, J.; Sripanidkulchai, K.; Suksamrarn, A.; Hu, Y.; Piyachuturawat, P.; Sripanidkulchai, B. Pharmacokinetics and organ distribution of diarylheptanoid phytoestrogens from *Curcuma comosa* in rats. *J. Nat. Med.* **2012**, *66*, 468–475. [CrossRef] [PubMed]
3. Winuthayanon, W.; Suksen, K.; Boonchird, C.; Chuncharunee, A.; Ponglikitmongkol, M.; Suksamrarn, A.; Piyachaturawat, P. Estrogenic Activity of Diarylheptanoids from *Curcuma comosa* Roxb. Requires Metabolic Activation. *J. Agric. Food Chem.* **2009**, *57*, 840–845. [CrossRef] [PubMed]
4. Suksamrarn, A.; Eiamong, S.; Piyachaturawat, P.; Byrne, L.T. A Phloracetophenone Glucoside with Choleretic Activity from *Curcuma comosa*. *Phytochemistry* **1997**, *45*, 103–105. [CrossRef]
5. Jurgens, T.M.; Frazier, E.G.; Schaeffer, J.M.; Jones, T.E.; Zink, D.L.; Borris, R.P.; Nanakorn, W.; Beck, H.T.; Balick, M.J. Novel nematocidal agents from *Curcuma comosa*. *J. Nat. Prod.* **1994**, *57*, 230–235. [CrossRef] [PubMed]
6. Matsumoto, T.; Nakamura, S.; Fujimoto, K.; Ohta, T.; Ogawa, K.; Yoshikawa, M.; Onishi, E.; Fukaya, M.; Matsuda, H. Structure of diarylheptanoids with antiallergic activity from the rhizomes of *Curcuma comosa*. *J. Nat. Med.* **2015**, *69*, 142–147. [CrossRef] [PubMed]
7. Wang, J.; Windbergs, M. Functional electrospun fibers for the treatment of human skin wounds. *Eur. J. Pharm. Biopharm.* **2017**, *119*, 283–299. [CrossRef] [PubMed]
8. Li, T.; Zhang, D.; Oo, T.N.; San, M.M.; Mon, A.M.; Hein, P.P.; Wang, Y.; Lu, C.; Yang, X. Investigation on the Antibacterial and Anti-T3SS Activity of Traditional Myanmar Medicinal Plants. *Evid. Based Complement. Altern. Med.* **2018**, *2018*, 2812908. [CrossRef] [PubMed]
9. Suksamrarn, A.; Ponglikitmongkol, M.; Wongkrajang, K.; Chindaduang, A.; Kittidanairak, S.; Jankam, A.; Yingyongnarongkul, A.B.E.; Kittipanumat, N.; Chokchaisiri, R.; Khetkam, P.; et al. Diarylheptanoids, new phytoestrogens from the rhizomes of *Curcuma comosa*: Isolation, chemical modification and estrogenic activity evaluation. *Bioorganic Med. Chem.* **2008**, *16*, 6891–6902. [CrossRef]
10. Yingngam, B.; Brantner, A.; Jinarat, D.; Kaewamatawong, R.; Rungseevijitprapa, W.; Suksamrarn, A.; Piyachaturawat, P.; Chokchaisiri, R. Determination of the marker diarylheptanoid phytoestrogens in *Curcuma comosa* rhizomes and selected herbal medicinal products by HPLC-DAD. *Chem. Pharm. Bull.* **2018**, *66*, 65–70. [CrossRef]
11. Intapad, S.; Saengsirisuwan, V.; Prasannarong, M.; Chuncharunee, A.; Suvitayawat, W.; Chokchaisiri, R.; Suksamrarn, A.; Piyachaturawat, P. Long-term effect of phytoestrogens from *Curcuma comosa* Roxb. on vascular relaxation in ovariectomized rats. *J. Agric. Food Chem.* **2012**, *60*, 758–764. [CrossRef]
12. Sodsai, A.; Piyachaturawat, P.; Sophasan, S.; Suksamrarn, A.; Vongsakul, M. Suppression by *Curcuma comosa* Roxb. of proinflammatory cytokine secretion in phorbol-12-myristate-13-acetate stimulated human mononuclear cells. *Int. Immunopharmacol.* **2007**, *7*, 524–531. [CrossRef] [PubMed]
13. Wiseman, H. The therapeutic potential of phytoestrogens. *Expert Opin. Investig. Drugs* **2000**, *9*, 1829–1840. [CrossRef] [PubMed]
14. Boonmee, A.; Srisomsap, C.; Chokchaichamnankit, D.; Karnchanatat, A.; Sangvanich, P. A proteomic analysis of Curcuma comosa Roxb. Rhizomes. *Proteome Sci.* **2011**, *9*, 43. [CrossRef] [PubMed]
15. Xu, F.; Nakamura, S.; Qu, Y.; Matsuda, H.; Pongpiriyadacha, Y.; Wu, L.; Yoshikawa, M. Structures of New Sesquiterpenes from *Curcuma comosa*. *Chem. Pharm. Bull.* **2008**, *56*, 1710–1716. [CrossRef]
16. Su, J.; Sripanidkulchai, K.; Hu, Y.; Chaiittianan, R.; Sripanidkulchai, B. Increased in situ intestinal absorption of phytoestrogenic diarylheptanoids from Curcuma comosa in nanoemulsions. *Off. J. Am. Assoc. Pharm. Sci.* **2013**, *14*, 1055–1062. [CrossRef] [PubMed]
17. Ahn, B.K.; Lee, S.G.; Kim, S.R.; Lee, D.H.; Oh, M.H.; Lee, M.W.; Choi, Y.W. Inclusion compound formulation of hirsutenone with beta-cyclodextrin. *J. Pharm. Investig.* **2013**, *43*, 453–459. [CrossRef]
18. Wani, T.A.; Shah, A.G.; Wani, S.M.; Wani, I.A.; Masoodi, F.A.; Nissar, N.; Shagoo, M.A. Suitability of different food grade materials for the encapsulation of some functional foods well reported for their advantages and susceptibility. *Crit. Rev. Food Sci. Nutr.* **2016**, *56*, 2431–2454. [CrossRef]
19. Radhakrishnan, S.; Lakshminarayanan, R.; Madhaiyan, K.; Barathi, V.A.; Lim, K.H.C.; Ramakrishna, S. Electrosprayed nanoparticles and electrospun nanofibers based on natural materials: Applications in tissue regeneration, drug delivery and pharmaceuticals. *Chem. Soc. Rev.* **2015**, *44*, 790–814.

20. STorres-Giner, R.; Pérez-Masiá, R.; Lagaron, J.M. A review on electrospun polymer nanostructures as advanced bioactive platforms. *Polym. Eng. Sci.* **2016**, *56*, 500–527. [CrossRef]
21. Bhattarai, R.S.; Bachu, R.D.; Boddu, S.H.S.; Bhaduri, S. Biomedical Applications of Electrospun Nanofibers: Drug and Nanoparticle Delivery. *Pharmaceutics* **2019**, *11*, 5. [CrossRef] [PubMed]
22. Papa, A.; Guarino, V.; Cirillo, V.; Oliviero, O.; Ambrosio, L. Optimization of Bicomponent Electrospun Fibers for Therapeutic Use: Post-Treatments to Improve Chemical and Biological Stability. *J. Funct. Biomater.* **2017**, *8*, 47. [CrossRef] [PubMed]
23. Yang, F.; Miao, Y.; Wang, Y.; Zhang, L.M.; Lin, X. Electrospun Zein/Gelatin Scaffold-Enhanced Cell Attachment and Growth of Human Periodontal Ligament Stem Cells. *Materials* **2017**, *10*, 1168. [CrossRef] [PubMed]
24. Nootem, J.; Chalorak, P.; Meemon, K.; Mingvanish, W.; Pratumyot, K.; Ruckthong, L.; Srisuwannaket, C.; Niamnont, N. Electrospun cellulose acetate doped with astaxanthin derivatives from Haematococcus pluvialis for in vivo anti-aging activity. *RSC Adv.* **2018**, *8*, 37151–37158. [CrossRef]
25. Zupacnic, S.; Baumgartner, S.; Lavric, Z.; Petelin, M.; Kristl, J. Local delivery of resveratrol using polycaprolactone nanofibers for treatment of periodontal disease. *J. Drug. Deliv. Sci. Technol.* **2015**, *30*, 408–416.
26. El-Refaie, K.; Bowlin, G.L.; Mansfield, K.; Layman, J.; Simpson, D.G.; Sanders, E.H.; Wnek, G.E. Release of tetracycline hydrochloride from electrospun poly (ethylene-co-vinylacetate), poly (lactic acid), and a blend. *J. Control. Release* **2002**, *81*, 57–64.
27. Chao, S.; Li, Y.; Zhao, R.; Zhang, L.; Li, Y.; Wang, C.; Li, X. Synthesis and characterization of tigecycline-loaded sericin/poly (vinyl alcohol) composite fibers via electrospinning as antibacterial wound dressings. *J. Drug. Deliv Sci. Technol.* **2018**, *44*, 440–447. [CrossRef]
28. Zhu, L.F.; Chen, X.; Ahmad, Z.; Li, J.S.; Chang, M.W. Engineering of Ganoderma lucidum polysaccharide loaded polyvinyl alcohol nanofibers for biopharmaceutical delivery. *J. Drug. Deliv Sci. Technol.* **2019**, *50*, 208–216. [CrossRef]
29. Ghorani, B.; Tucker, N. Fundamentals of electrospinning as a novel delivery vehicle for bioactive compounds in food nanotechnology. *Food Hydrocoll.* **2015**, *51*, 227–240. [CrossRef]
30. Figueroa-Lopez, K.J.; Castro-Mayorga, J.L.; Andrade-Mahecha, M.M.; Cabedo, L.; Lagaron, J.M. Antibacterial and barrier properties of gelatin coated by electrospun polycaprolactone ultrathin fibers containing black pepper oleoresin of interest in active food biopackaging applications. *Nanomaterials* **2018**, *8*, 199. [CrossRef]
31. Mogosanu, G.D.; Grumezescu, A.M. Natural and synthetic polymers for wounds and burns dressing. *Int. J. Pharm.* **2014**, *463*, 127–136. [CrossRef] [PubMed]
32. Maytal, F.; Zilberman, M. Drug delivery from gelatin-based systems. *Expert Opin. Drug Deliv.* **2015**, *12*, 1547–1563.
33. Nagarajan, S.; Soussan, L.; Bechelany, M.; Teyssier, C.; Cavailles, V.; Pochat-Bohatier, C.; Miele, P.; Kalkura, N.; Janot, J.M.; Balme, S. Novel biocompatible electrospun gelatin fiber mats with antibiotic drug delivery properties. *J. Mater. Chem. B* **2016**, *4*, 1134–1141. [CrossRef]
34. Li, H.; Wang, M.; Williams, G.R.; Wu, J.; Sun, X.; Lv, Y.; Zhu, L.M. Electrospun gelatin nanofibers loaded with vitamins A and E as antibacterial wound dressing materials. *RSC Adv.* **2016**, *6*, 50267–50277. [CrossRef]
35. Sutjarit, N.; Sueajai, J.; Boonmuen, N.; Sornkaew, N.; Suksamrarn, A.; Tuchinda, P.; Zhu, W.; Weerachayaphorn, J.; Piyachaturawat, P. *Curcuma comosa* reduces visceral adipose tissue and improves dyslipidemia in ovariectomized rats. *J. Ethnopharmacol.* **2018**, *215*, 167–175. [CrossRef]
36. Prashanth, N.; Bhavani, N.L. Phytochemical analysis of two high yielding *Curcuma longa* varieties from Andhra Pradesh. *Int. J. Life Sci. Biotechnol. Pharm. Res.* **2013**, *2*, 103–108.
37. Devi, K.P.; Suganthy, N.; Kesika, P.; Pandian, S.K. Bioprotective properties of seaweeds: In vitro evaluation of antioxidant activity and antimicrobial activity against food borne bacteria in relation to polyphenolic content. *BMC Complement. Altern. Med.* **2008**, *8*, 1–11. [CrossRef]
38. Masamoto, Y.; Ando, H.; Murata, Y.; Shimoishi, Y.; Tada, M.; Takahata, K. Mushroom tyrosinase inhibitory activity of esculetin isolated from seeds of *Euphorbia lathyris* L. *Biosci. Biotechnol. Biochem.* **2003**, *67*, 631–634. [CrossRef]

© 2019 by the authors. Licensee MDPI, Basel, Switzerland. This article is an open access article distributed under the terms and conditions of the Creative Commons Attribution (CC BY) license (http://creativecommons.org/licenses/by/4.0/).

Article

Hydrogel Nanofibers from Carboxymethyl Sago Pulp and Its Controlled Release Studies as a Methylene Blue Drug Carrier

Nafeesa Mohd Kanafi [1], Norizah Abdul Rahman [1,2,*], Nurul Husna Rosdi [1], Hasliza Bahruji [3] and Hasmerya Maarof [4]

1. Department of Chemistry, Faculty of Science, Universiti Putra Malaysia, 43400 UPM Serdang, Malaysia; nafeesakanafi@yahoo.com (N.M.K.); nurulhusnarosdi73@gmail.com (N.H.R)
2. Laboratory of Materials Processing and Technology, Institute of Advanced Technology, Universiti Putra Malaysia, 43400 UPM Serdang, Malaysia
3. Centre of Advanced Material and Energy Science, University Brunei Darussalam, Jalan Tungku Link, BE 1410, Brunei Darussalam; hasliza.bahruji@ubd.edu.bn
4. Department of Chemistry, Faculty of Science, Universiti Teknologi Malaysia, 81310 UTM Johor Bahru, Malaysia; hasmerya@kimia.fs.utm.my
* Correspondence: a_norizah@upm.edu.my

Received: 12 May 2019; Accepted: 3 June 2019; Published: 15 June 2019

Abstract: The potential use of carboxymethyl sago pulp (CMSP) extracted from sago waste for producing hydrogel nanofibers was investigated as a methylene blue drug carrier. Sago pulp was chemically modified via carboxymethylation reaction to form carboxymethyl sago pulp (CMSP) and subsequently used to produce nanofibers using the electrospinning method with the addition of poly(ethylene oxide) (PEO). The CMSP nanofibers were further treated with citric acid to form cross-linked hydrogel. Studies on the percentage of swelling following the variation of citric acid concentrations and curing temperature showed that 89.20 ± 0.42% of methylene blue (MB) was loaded onto CMSP hydrogel nanofibers with the percentage of swelling 4366 ± 975%. Meanwhile, methylene blue controlled release studies revealed that the diffusion of methylene blue was influenced by the pH of buffer solution with 19.44% of MB released at pH 7.34 within 48 h indicating the potential of CMSP hydrogel nanofibers to be used as a drug carrier for MB.

Keywords: carboxymethyl sago pulp; controlled release; electrospinning; hydrogel; nanofiber

1. Introduction

Sago palm (*Metroxylon sago*) is largely grown in tropical lowland forest and swamp areas. Sarawak, Malaysia is recognized as the largest sago-growing area with an estimated 54,000 hectares in 2013 [1,2]. Sago starch is the main product of sago palm, widely exported to the Peninsular Malaysia, Japan, Singapore and other countries with a total amount of 48,000 tons in 2013 [3]. The extraction of sago starch involves removing the bark of sago palm followed by rasping and sieving of the stem to produce starch slurry. The resulting slurry is then put to further settling, washing and drying processes to produce purified starch. The process also produces a fibrous sago waste by-product, with approximately 7 tons of production from a daily single processing mill [2]. Sago waste is a light brown lignocellulosic biomass comprising cellulose, hemicellulose and lignin. During the pre-treatment process, which is bleaching, hemicellulose and lignin are solubilized and degraded, thus resulting in sago pulp rich in cellulose [1,4]. Cellulose is a long and linear polysaccharide polymer that consists of glucose units that linked via beta 1,4-glycosidic bond. Studies on the production of functionalized cellulose derivatives, such as carboxymethylcellulose [2], hydroxypropyl methylcellulose (HPMC) and methylcellulose [5] have revealed their potential use as a drug carrier in pharmaceutical industries [2,4].

Hydrogel is often produced using chemical or physical cross-linking methods to form a three-dimensional polymeric structure capable of retaining water or biological fluid beyond its dry weight without dissolution in water [6–8]. The ionic groups within the hydrogel structure respond towards external changes such as pH, ionic strength and temperature which consequently influence the uptake and release of fluid [9]. It has been extensively studied as an excellent material for biomedical applications due to its softness, super-absorbency, biodegradability, biocompatibility, hydrophilicity and similarity with extracellular matrix (ECM) [10]. As the hydrogel structure can simulate the ECM, it has been widely applied in drug delivery and wound dressing from various synthetic and natural materials [11,12]. Furthermore, hydrogel has a low chance of triggering negative immune responses in the human body as it has a low interfacial tension with biological fluids that reduce cell adhesion. Among the various types of stimuli responsive hydrogels, pH sensitive hydrogels are the most studied, especially in drug delivery and wound dressing applications [10].

Carboxymethyl sago pulp (CMSP) is also known as carboxymetyl cellulose. It is a functionalized sago pulp with the presence of carboxymethyl groups bound to the cellulose backbone. The hydroxyl groups bound to carboxymethyl groups are responsible for its adsorption behavior [13]. The preparation of carboxymethylated cellulose from various agricultural-by-products through the carboxymethylation process in alkaline conditions has been well documented in the literature [2,13–17]. CMSP hydrogel is sensitive towards the environment and tends to adsorb a large amount of fluid in various pH media, leading to swelling on its structure [18]. The swelling of CMSP hydrogel reported by Pushpamalar et al. showed that the hydrogel has a high swelling ratio at pH 7 [18]. Furthermore, CMSP hydrogel in the form of microdiscs and microbeads loaded with ciprofloxacin and 5-aminosalicylic acid showed its potential application as a drug carrier due to its ability to release a drug at different pH [19,20].

Designing of nanomaterials with low density, large surface area, high pore volume and tight pore size has been the main focus of research in biomedical fields [21,22]. Electrospinning is a highly versatile method for producing nanofibers using natural or synthetic polymers, which can be used in drug and therapeutic agent deliveries, wound dressings and in the tissue engineering field [11,22,23]. In recent years, many studies on nanofibrous cellulose derivatives, including cellulose acetate nanofibers, have been applied in various applications, including biomedical [24]. In thist study, the production of CMSP hydrogel nanofibers from sago waste was investigated for controlled drug release applications. The CMSP nanofibers were prepared by incorporation with poly(ethylene oxide) (PEO) via an electrospinning technique with relatively low voltage used (20 kV) compared to electrospun carboxymethyl cellulose (CMC)/PEO fibers as reported previously [5,25]. To the best of the authors' knowledge, no study has been performed for CMSP/PEO nanofibers fabricated using electrospinning techniques followed by treatment with citric acid, a cross-linker to form hydrogel. As an electro-spinnable and biocompatible polymer, PEO was blended with CMSP to facilitate the formation of nanofibers. The controlled release of methylene blue (MB) from CMSP hydrogel nanofibers was investigated at different pH solutions. MB acts as a drug model with many uses in clinical medicine including antimalarial agents, in the treatment of early Alzheimer's and methemoglobinemia [26–28]. Furthermore, MB is commonly used as preliminary test for hydrogel-controlled release and adsorption studies [29–31].

2. Materials and Methods

2.1. Materials

The CMSP (DS = 0.6) from sago waste was prepared based on the method reported by Pushpamalar et al. (2006) [2]. Poly(ethylene oxide) (PEO), citric acid and sodium monochloroacetate were purchased from Sigma Aldrich (St. Louis, MO, USA). Sodium hydroxide was bought from Bio Basic Canada Inc. (Markham, ON, Canada). Isopropanol, methanol and nitric acid were purchased from J. Kollin Chemicals (UK), while glacial acetic acid was purchased from J.T. Baker (Phillipsburg, NJ, USA).

Methylene blue was obtained from Friendemann Schmidt (Parkwood, WA, USA) and distilled water was used throughout the study. All chemicals were used without further purification.

2.2. Carboxymethylation of Sago Pulp

To start with, 10.0 g of dried ground sago pulp was added into the dissolution tester followed by the addition of 200 mL of isopropanol with continuous stirring at 60 rpm for 30 min. Then, 20 mL of 25% (wt/vol) sodium hydroxide solution was added drop-wise into the mixture and stirred for an hour at room temperature. Following this, 12.0 g of sodium monochloroacetate was added into the reaction mixture for carboxymethylation reaction and the speed of the stirrer was increased to 100 rpm. The resulting mixture was stirred for another 30 min and heated at 45 °C for 3 h with continuous stirring. The mixture was cooled to room temperature and filtered to obtained carboxymethyl sago pulp (CMSP) residue. The residue was thoroughly washed with 100% methanol and treated with glacial acetic acid before drying to form CMSP powder. The degree of substitutions (DS) of the prepared CMSP was determined by potentiometric back titration [2]. Firstly, CMSP was treated using nitric acid. Then, 0.5 g of acid CMSP was weighed into the Erlenmeyer flask followed by the addition of 100 mL of distilled water and stirred. Subsequently, 25 mL of 0.5 M sodium hydroxide was added, stirred and the solution was boiled for approximately 15 min until dissolved. Finally, 1 drop of phenolphthalein as an indicator was added into the heated solution and titrated with 0.3 N of hydrochloric acid.

2.3. Preparation of CMSP/PEO Hydrogel Nanofibers

The CMSP nanofibers were prepared using the electrospinning method. PEO was added to CMSP to facilitate the formation of nanofibers. The CMSP/PEO solution was prepared by dissolving the CMSP/PEO in water. CMSP and PEO were weighed at various ratios of CMSP to PEO prior to dissolution in water. The mixture was stirred in a beaker at room temperature to form homogenous solution. The CMSP/PEO solution was filled in a 5 mL syringe fitted with a metal needle for the electrospinning process. The applied voltage and the tip-to-collector distance were kept at 20 kV and 15 cm, respectively. The parameters used to optimize CMSP nanofibers production were varied with (a) the ratio of CMSP to PEO in the range of 1:0, 3:1, 1:1 and 1:3; (b) the concentration of CMSP/PEO solution from 4%, 6%, 8% and 10% (wt/vol%), and (c) the flow rate of CMSP/PEO solution during the electrospinning at 0.6 mL/h, 0.8 mL/h, 1.0 mL/h and 1.2 mL/h. The electrospinning process was conducted at room temperature with a horizontal setup while humidity was kept below 60% using silica gel to reduce humidity in the electrospinning box. The optimized electrospinning conditions were used to prepare CMSP nanofibers for preparing hydrogel nanofibers. The effects of citric acid concentration and the curing temperature for production of CMSP hydrogel nanofibers were investigated with percentage of citric acid varied at 23%, 29%, 33% and 38%, as well as curing temperatures at 60 °C, 70 °C, 80 °C and 90 °C. The curing period was kept constant at 450 min.

2.4. Swelling Studies of CMSP Hydrogel Nanofiber

The swelling studies were carried out on the optimized hydrogel nanofibers. The cross-linked dry nanofibers were weighed (Wd) before immersion in 15 mL of distilled water for 24 h. The weight of nanofibers hydrogel (Ws) was measured following 24 h immersion and the percentage of swelling was calculated according to Equation (1) [8,32]. The experiments were carried out in triplicate to obtain an average value. The balance used was Mettler Toledo with 4 decimal points. Later, the swelled optimized hydrogel nanofibers were dried in an oven overnight to constant weight at 50 °C before drug loading.

$$Q\,(\%) = \frac{Ws - Wd}{Wd} \times 100 \qquad (1)$$

where Q = percentage of swelling, Ws = weight of nanofibers hydrogel, Wd = weight of the initial nanofibers.

2.5. Drug Controlled Release Studies

Methylene blue (MB) was loaded into dried hydrogel nanofibers by immersion in 100 ppm of methylene blue solution in a petri dish for 24 h. The petri dish was covered with aluminum foil to prevent degradation of MB with light. After 24 h, the MB-loaded hydrogel was washed with distilled water and excess MB solution on the surface of the hydrogel was removed using filter paper. The efficiency of methylene blue entrapment was determined by analyzing the concentration of MB before and after immersion with nanofibers. Concentration of methylene blue was determined using UV-vis spectrophotometer (Shimadzu, Kyoto, Japan). Meanwhile, the MB entrapment efficiency was calculated based on Equation (2) [33,34]:

$$Entrapement\ efficiency\ (EE)\ (\%) = \frac{(total\ initial\ concentration\ of\ MB - free\ MB\ concentration)}{total\ initial\ concentration\ of\ MB} \times 100 \quad (2)$$

In controlled release studies, MB-loaded hydrogel was immersed in phosphate-buffer saline (PBS) with pH 7.34 at room temperature for 48 h. Further, 3 mL aliquot was withdrawn periodically and similar volumes of fresh PBS solution was replaced. The collected samples were measured by determining the absorption of light at 664 nm using a UV-vis spectrophotometer. The procedures were repeated at pH 1.02, 4.0 and 8.0 of buffer solutions. The cumulative percentage of MB release was calculated based on Equation (3) [35]:

$$Cumulative\ percentage\ release\ (\%) = \frac{vol.\ of\ sample\ withdrawn\ (ml)}{bath\ volume\ (ml)} \times P\ (t-1) + Pt \quad (3)$$

where

Pt = Percentage release at time t (minutes)

$P\ (t-1)$ = Percentage release previous to 't'

2.6. Fourier Transform Infrared (FTIR) Spectroscopy

Fourier transform infrared (FTIR) spectroscopy was used to characterize the presence of functional groups in the pre-treated sago pulp and CMSP. FTIR spectra of the materials were obtained using Perkin Elmer Spectrum 100 Fourier Transform Infrared Spectrophotometer (Perkin Elmer, Selton, CT, USA). Scanning was carried out in the range 4000–280 cm^{-1}. Attenuated Total Reflection (ATR) [6] was utilized as the sampling technique.

2.7. Ultraviolet-Visible (UV-Vis) Spectrophotometer

A UV-visible (UV-vis) spectrophotometer was used to determine the concentration of MB. Ultraviolet measurements were performed using Shimadzu UV-visible spectrophotometer (Shimadzu, Kyoto, Japan). The wavelength number range to detect the absorption was set at 400–800 nm and the MB λ_{max} = 664 nm.

2.8. Scanning Electron Microscopy (SEM)

Scanning electron microscopy was used to determine surface morphology of nanofibers produced using the electrospinning method and hydrogel nanofibers following the immersion in water and MB solution. For swelling nanofibers, the samples were freeze-dried before SEM analysis. All the nanofibers were observed using the scanning electron microscopy (SEM) model JEOL JSM 6400 (JEOL, Tokyo, Japan) attached with energy-dispersive X-ray (EDX). The samples were sputter coated with a thin layer of gold before being observed under the SEM at a voltage of 15 kV. More than 100 of the nanofibers were randomly measured from SEM images by using the image J program (Version 1.52o, SpartanCoders, Softonic International S.A.) to measure the mean diameter of the fibers.

3. Results and Discussion

3.1. Fourier Transform Infrared Spectroscopy Studies

Infrared analysis was carried out on the pre-treated sago pulp and CMSP as shown in Figure 1. The pre-treated sago pulp showed a broad O–H stretching vibration band at 3600–3000 cm^{-1} accompanied with the OH bending vibration of the absorbed water band at 1623 cm^{-1} [36]. However, in CMSP, the intensity of the O–H stretching vibration was reduced with the OH bending vibration band shifts toward lower wavenumber at 1596 cm^{-1}. The shift of the peak is associated with the COO stretching vibration in CMSP. The features provide evidence on the substitution of OH with the carboxymethyl group in CMSP following the carboxymethylation reaction. The CMSP also demonstrated a strong absorbance band at 1400 cm^{-1} in comparison to sago pulp which corresponded to the –CH$_2$ scissoring vibration. The sago pulp and CMSP also revealed absorbance bands at 1030–1020 cm^{-1} which corresponded to the –CH–O–CH$_2$ stretching vibration of the ether group [17]. In sago pulp, the ether group originated from the linkage of 1,4-β-D-glucoside group (>CH–O–CH<) in the cellulose molecule. In CMSP however, the absorbance band of ether also originated from the ether group of the carboxymethylation of sago pulp, in which the two types of ether were overlapped in the CMSP [2,17,18].

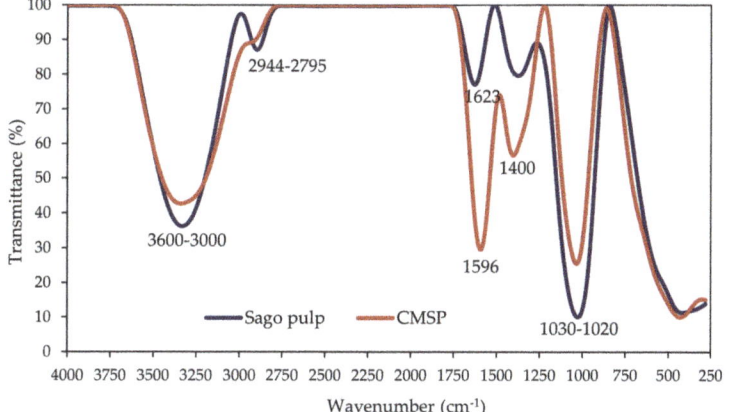

Figure 1. Fourier transform infrared (FTIR) spectra pattern of pre-treated sago pulp and carboxymethyl sago pulp (CMSP).

3.2. Optimization of Electrospinning Parameters of CMSP Nanofibers

Optimization of the electrospinning parameters was carried out by varying the weight ratio of CMSP to PEO, the concentration of CMSP/PEO solutions and the CMSP/PEO solution flow rates. The applied voltage was fixed at 20 kV while the tip-to-collector distance was fixed at 15 cm. The electrospinning process was conducted at room temperature while the humidity was kept below 60%.

3.2.1. Effects of Weight Ratio of CMSP to PEO

CMSP is a natural polymer that requires synthetic polymer to assist the electrospinning process. In this study, CMSP was mixed with a biocompatible polymer, PEO, to facilitate the electrospinning process by increasing the elasticity of the CMSP solution [37]. Figure 2 shows the morphology and the average diameter of nanofibers at three different weight ratios of CMSP to PEO. The morphology of the resulting fibers had gradually transformed from microbeads to uniform fiber structure following the variation of CMSP to PEO ratios from 3:1 to 1:3. There was no SEM image to study the morphology of ratio 1CMSP:0PEO because CMSP cannot be electrospun alone. It needs to be blended with electro-spinnable polymer, PEO, to allow CMSP to be electrospun into nanofibers. In the presence of high CMSP content as in 3CMSP:1PEO ratio, the electrospinning process produced a microbead structure with only a few nanofibers as observed from the SEM analysis. At equal ratio of CMSP and PEO, the electrospinning of the mixtures produced nanofibers with wide fiber diameter distribution between 100–1000 nm. However, the presence of microbead structures was still visible from SEM images. At 1CMSP:3PEO ratio, the formation of nanofibers with narrow diameter distribution were observed with average fiber diameter in the range of 51–250 nm. The optimization of the electrospinning process continued using the CMSP:PEO ratios of 1:1 and 1:3, due to significantly less microbead structures produced.

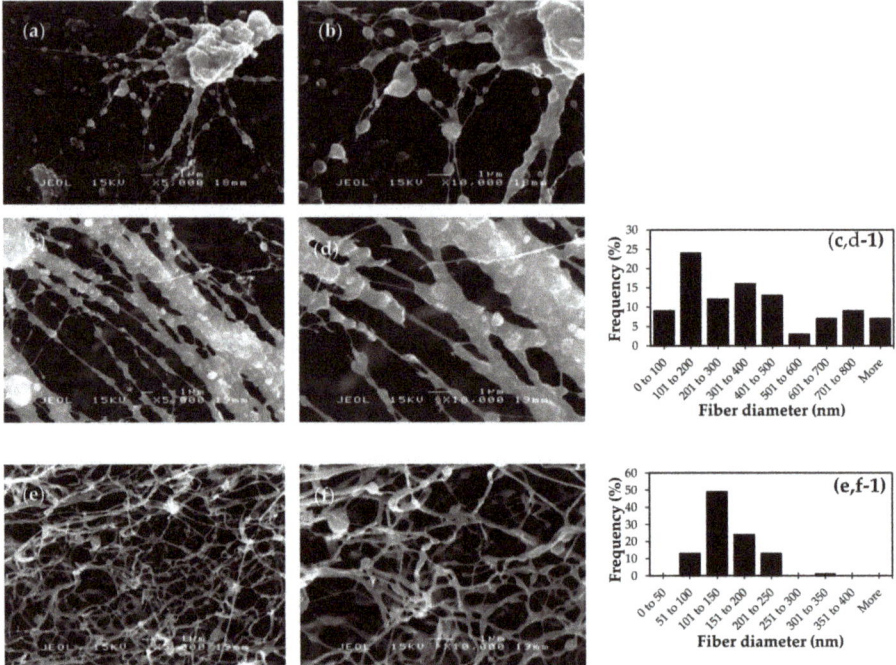

Figure 2. SEM images (**a–f**) and the corresponding fiber diameter distribution ((**c,d-1**)–(**e,f-1**)) of CMSP nanofibers at variation of CMSP to poly(ethylene oxide) (PEO) weight ratios (**a,b**) 3:1, (**c,d**) 1:1 and (**e,f**) 1:3 at fixed 6% solution concentration, 20 kV applied voltage, 15 cm of tip-to-collector distance and 0.8 mL/h of flow rate.

3.2.2. Effects of Concentration of CMSP/PEO Solution

The morphology of nanofibers is significantly affected by the concentration of polymer solution during the electrospinning process [38,39]. In this study, electrospinning was carried out at four different concentrations of CMSP/PEO solutions ~4%, 6%, 8% and 10% (wt/vol%) with the ratio of CMSP to PEO kept at 1:1 and 1:3. SEM images shown in Figure 3 reveal the concentration of solution that significantly affected the formation of nanofibers. At 4% of solution concentration, the presence of microbead structures was observed together with the nanofibers, with average diameter ranging from 100–200 nm. Increasing the concentration had improved the formation of nanofibers with the absence of microbead structures. At 10% concentration, nanofibers obtained at 1:3 and 1:1 ratios showed uniform nanofibers morphology with no visible formation of microbeads. High concentration of polymer in the electrospinning jet significantly enhanced the interaction between polymer chains in solution, which led to a greater resistance of the solution against the pulling force by electrical charges [40]. Nanofibers with narrow mean diameter of 214 ± 62 nm was produced at 10% concentration using 1CMSP:1PEO ratio, which was used for further optimization studies of CMSP/PEO electrospinning.

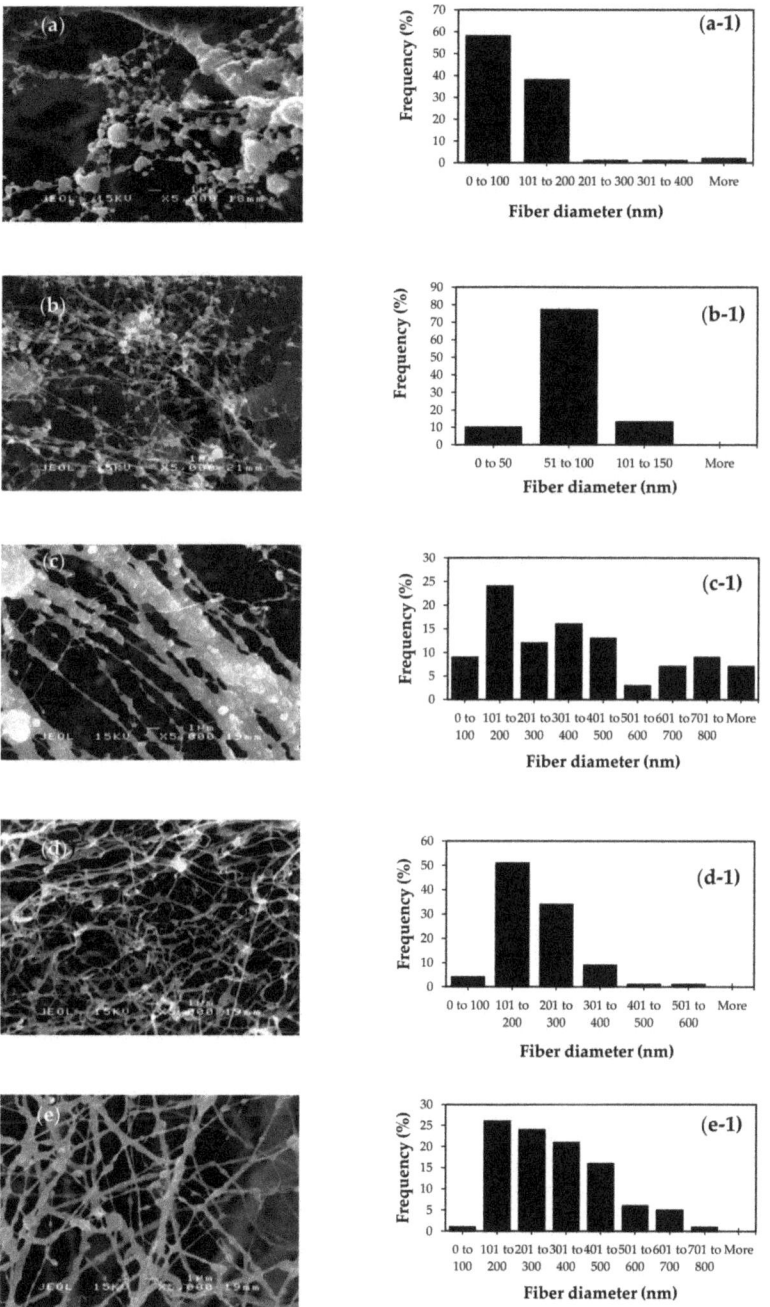

Figure 3. *Cont.*

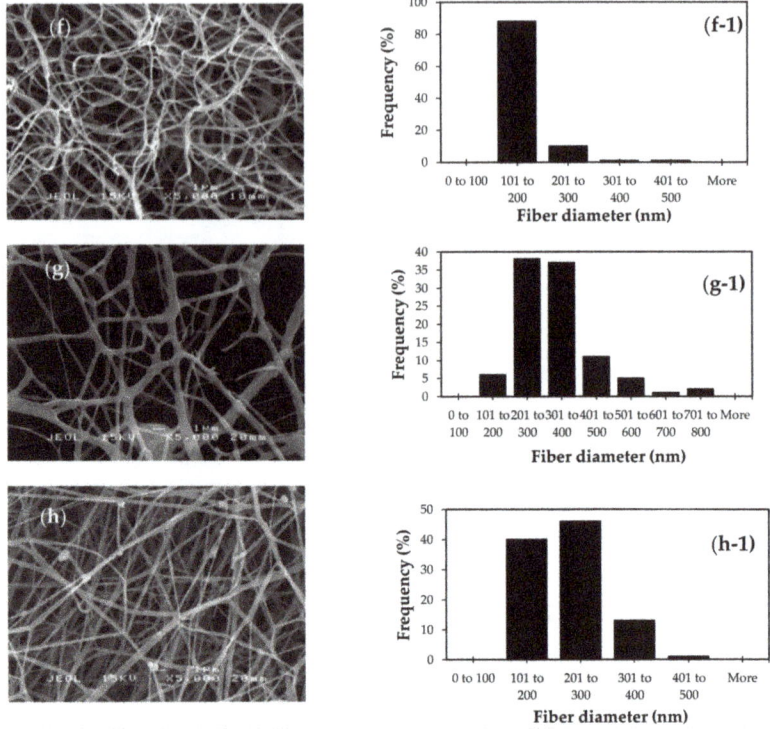

Figure 3. SEM images (**a–h**) and the corresponding fiber diameter distribution ((**a-1**)–(**h-1**)) of CMSP/PEO nanofibers at different concentration of CMSP to PEO solution and ratio of CMSP to PEO (**a**) 4% (ratio 1:1), (**b**) 4% (ratio 1:3), (**c**) 6% (ratio 1:1), (**d**) 6% (ratio 1:3), (**e**) 8% (ratio 1:1), (**f**) 8% (ratio 1:3), (**g**) 10% (ratio 1:1) and (**h**) 10% (ratio 1:3). Electrospinning parameters were kept constant at 20 kV for applied voltage, 15 cm of tip-to-collector distance and 0.8 mL/h for flow rate.

3.2.3. Effects of Flow Rate of Syringe Pump

The influence of flow rate during the electrospinning process was investigated by variation at 0.6, 0.8, 1.0, and 1.2 mL/h and the resulting nanofibers are shown in Figure 4. Nanofibers with a diameter of 280 ± 89 nm were produced at 0.6 mL/h. When the flow rate was increased to 1.2 mL/h, the diameter of nanofibers significantly increased with the maximum size of 355 ± 185 nm obtained at 1.2 mL/h rate. Increasing the flow of solution reduced the time for nanofibers to dry and therefore prompted integration of nanofibers to form a much larger structure [29,41,42]. The nanofibers produced at 0.6 mL/h have relatively uniform structures, although the diameter of resulting nanofibers was significantly larger than nanofibers produced at 0.8 and 1.0 mL/h. Therefore, the flow rate of 0.6 mL/h was considered as the optimum rate for fabrication of CMSP nanofibers. For investigating the degree of swelling, the optimized parameters of 10% solution concentration, 1CMSP:1PEO ratio, 0.6 mL/h flow rate, with the constant 20 kV of applied voltage and 15 cm of tip-to-collector distance, were used to produce hydrogel nanofibers.

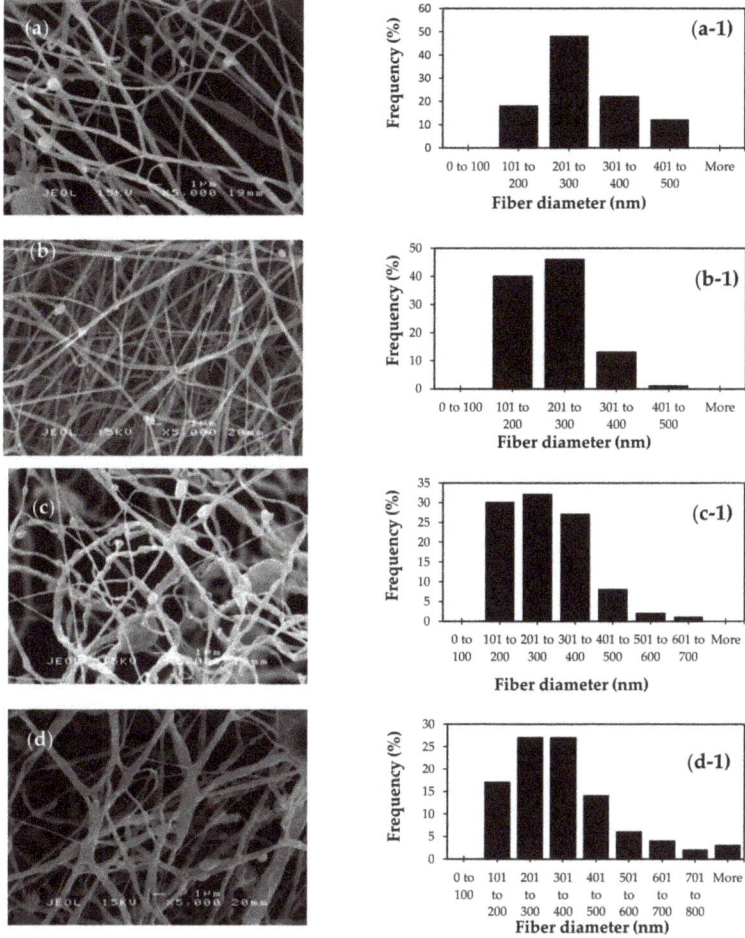

Figure 4. SEM images (**a**–**d**) and the corresponding fiber diameter distribution ((**a-1**)–(**d-1**)) of CMSP nanofibers produced at flow rate of (**a**) 0.6 mL/h, (**b**) 0.8 mL/h, (**c**) 1.0 mL/h and (**d**) 1.2 mL/h at fixed 10% concentration, 1:1 ratio, 20 kV of applied voltage and 15 cm tip-to-collector distance.

3.3. Swelling Studies of CMSP Hydrogel Nanofibers

Investigation on the swelling percentage of CMSP nanofibers was determined using gravimetric method at a variation of citric acid concentrations and curing temperatures. The duration of curing was kept at 450 min. The percentage of swelling will determine the amount of fluid that can penetrate into the polymeric network, which provides an understanding on the amount of drugs that can retained within the nanofibers structure [43].

3.3.1. Effects of Percentage of Citric Acid

Figure 5a shows the relationship between the percentages of swelling upon increasing the concentrations of citric acid. At 23% of citric acid concentration, the percentage of swelling showed a maximum value of 10,029 ± 449%. However, the percentage of swelling has significantly reduced to 1491 ± 126% at 29% of citric acid concentration. Although the nanofibers showed high percentage of swelling at 23% of citric acid concentration, the hydrogel nanofibers were partially dissolved after

24 h. Citric acid is a cross-linker that bridges two polymeric chains to form a network that will hold water without initiating dissolution. Figure 6 shows the schematic diagram of a cross-linking reaction between hydroxyl groups from CMSP and citric acid molecules. Here, CMSP plays a dominant role in the cross-linking process compared to PEO. PEO has a long linear chain, has only one hydroxyl group at each end of the chain, while CMSP has much more hydroxyl groups. Low concentrations of citric acid produced a small amount of cross-linked networks relative to the amount of hydroxyl groups available for the interaction with water in the hydrogel. In the presence of a large volume of water, the cross-linked network collapsed and consequently increased the dissolution in water. This occurred for hydrogel nanofibers with 23% of citric acid, but not at the higher percentage of citric acid. Moreover, the reason why 23% was chosen as the starting point is because of low percentages lead to total dissolution of hydrogel nanofibers in water. The percentage of swelling for hydrogel nanofibers was decreased when the concentration of citric acid increased to 29%. At high concentrations of citric acid, a high percentage of cross-linked polymeric networks was produced, which significantly improved the ability to retain water molecules in the structure. However, a high percentage of cross-linking also reduced the amount of free hydroxyl groups in the structure, therefore limiting the amount of water that can be accommodated within the network [7,44,45].

3.3.2. Effects of Curing Temperatures

Curing is a process where the nanofibers are treated in an oven at certain periods of time for a cross-linking reaction to occur. The percentage of swelling was further investigated by increasing the curing temperatures of the hydrogel nanofibers at 60 °C, 70 °C, 80 °C and 90 °C. The results in Figure 5b show the highest percentage of swelling at 4366 ± 975% when curing temperatures was at 80 °C. Increasing the temperature from 60 °C to 80 °C gradually improved the percentage of swelling due to the increment of the cross-linked network within the structure. When temperature increased for a constant time duration, the molecules of CMSP, PEO and citric acid moved quicker, while the energy of the molecules increased. This led to a greater frequency of successful collisions between reactant molecules and the greater the rate of a cross-linking reaction. However at 90 °C, the percentage of swelling was slightly decreased, suggesting a large amount of cross-linking between the polymers produced and caused a rigid cross-linked network. Optimization of the amount of citric acid and the curing temperature on the degree of swelling provides an insight into the ability of hydrogel to adsorb and hold methylene blue within the polymeric network. The optimum CMSP/PEO hydrogel nanofibers (10% concentration, 1:1 ratio, 29% of citric acid, 80 °C of curing temperature and 450 min of curing time) was prepared to proceed with the drug release study.

In this swelling study, a traditional and widely used method of measuring the percentage of swelling is based on the weight of the hydrogel. Inaccuracy of measuring the percentage of swelling may occur due to difficulty in handling swelling hydrogel nanofibers as they have a delicate structure. Structural destruction of hydrogel might occur during weight measurement because swelling of hydrogel nanofibers having low mechanical properties. Extra precaution was taken in this study to ensure accurate result. Recently, a new and more efficient method had been found by Tavakoli et al. (2019) to overcome all the limitations and errors in traditional methods which is known as the fluorescence intensity method (FL). This new method can be employed for obtaining accurate measurements of percentage of swelling hydrogels, especially delicate nanofibers [46].

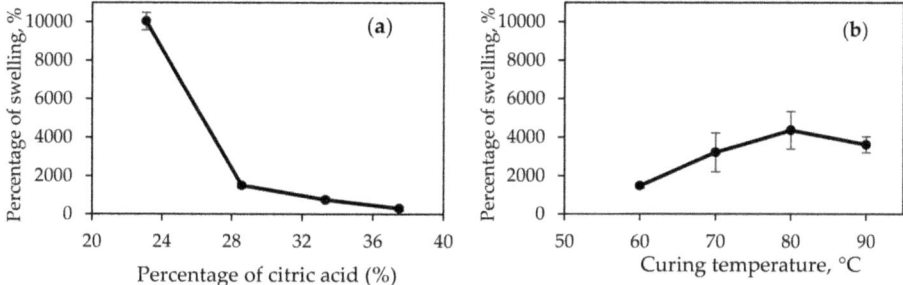

Figure 5. Plot of percentage of swelling at (**a**) different percentage of citric acid (23%, 29%, 33% and 38%) at fixed 60 °C and 450 min of curing; and (**b**) different curing temperature (60 °C, 70 °C, 80 °C and 90 °C) but at fixed 29% citric acid, 450 min of curing.

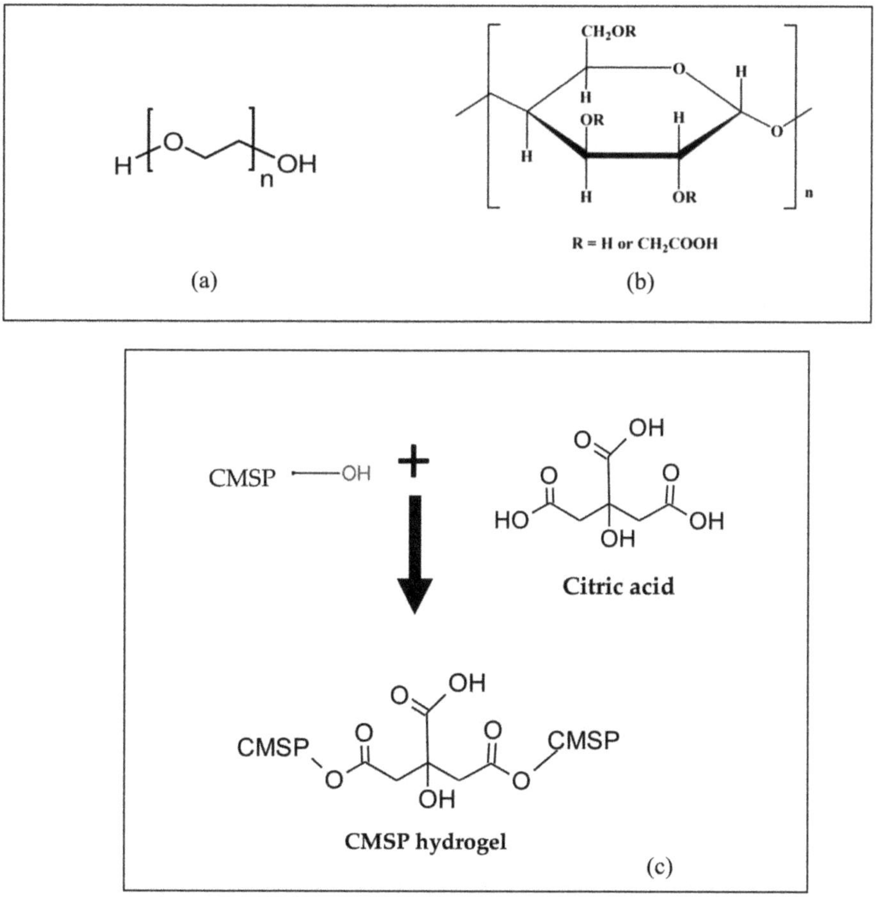

Figure 6. (**a**) PEO structure (**b**) CMSP structure (**c**) Proposed mechanism of cross-linking reaction between cellulose (CMSP) and citric acid.

3.4. Morphology Studies

SEM images of the cross-linked CMSP hydrogel nanofibers following immersion in water in swelling studies and adsorption with MB are shown in Figure 7. The morphology of CMSP nanofibers hydrogel was significantly different compared to CMSP nanofibers as shown in Figure 4. The SEM image in Figure 7a shows the CMSP hydrogel nanofibers that were still in nanofibrous form after the cross-linking reaction. This proved that a cross-linking reaction does not affect the morphology of the CMSP nanofibers. The SEM analysis in Figure 7b,c demonstrates swollen nanofibers with the appearance of porous structures. The expansion of the fiber diameter and the formation of porous structures proved that the absorption of water into the hydrogel network produced a high percentage of swelling of the CMSP nanofibers. The nanofibers were cross-linked to form a polymeric network with voids to hold water. The SEM images in Figure 7d,e illustrate the morphology of hydrogel nanofibers following immersion with MB was different than immersion with water, whereby the nanofibers showed bigger enlargement with less pores and void spaces within the hydrogel network.

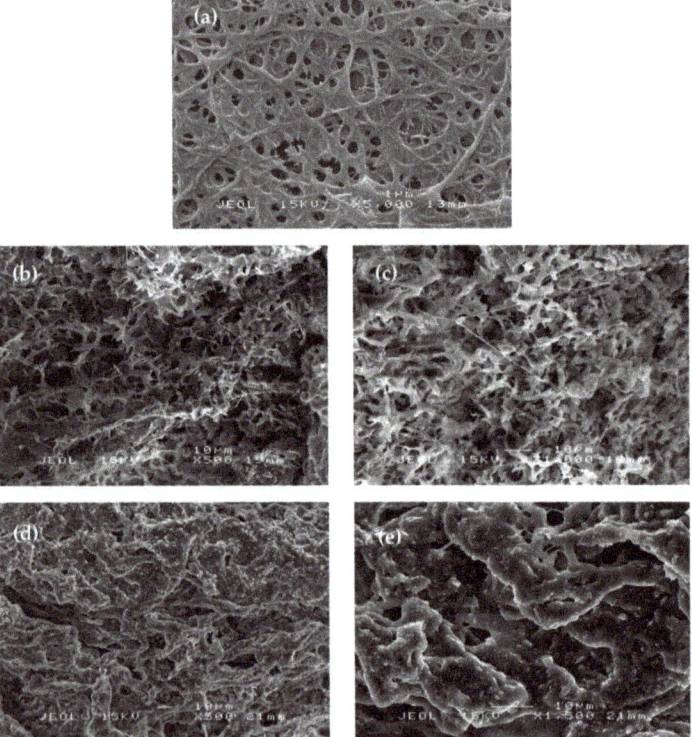

Figure 7. SEM images of (**a**) cross-linked CMSP hydrogel nanofibers; (**b,c**) swollen CMSP hydrogel nanofibers; and (**d,e**) methylene blue (MB)-loaded CMSP hydrogel nanofibers.

3.5. Drug Release Studies

The potential of CMSP hydrogel nanofibers to be used as a drug carrier was investigated on the ability to release the adsorbed MB at various pHs. Investigations on pH variation is important to simulate the physiological environments along the human gastrointestinal (GI) tract starting with the saliva at pH 5–6, stomach at pH 1–3, small intestine at pH 6.6–7.5 and colon at pH 6.4–7.0 [10,47,48]. When necessary, the drug release ability at pH 7.15–8.9 is crucial to heal chronic wounds [10,49].

However, the delicate structure of hydrogel nanofibers may be unsuitable for oral administration. Nevertheless, it can be further applied as wound dressings in wound treatments. Drug release from polymers occurs when a liquid medium penetrates the polymeric network leading to the release of the retained drug by diffusion. Figure 8 shows the MB release profiles from hydrogel nanofibers at different pH of buffer solutions. In general, the CMSP shows a high degree of MB releases at high pH, with the highest cumulative percentage after 2880 min obtained at pH 8.0 (21.21%). Within 30 min into the investigation, the acceleration of MB diffusion was observed in all pH media with similar cumulative percentage releases of ~5% suggesting the diffusion of MB from CMSP hydrogel nanofibers trapped on the surface of the nanofibers. The initial diffusion of the surface trapped MB also occurred at a similar rate which suggests that the process was not influenced by the pH of the solution. Monitoring the concentration of MB after 6 h showed a rapid diffusion of MB when the pH of buffer solution was 7.34, and the rates were slowly reduced as the pH decreases to 1.2. At pH 1.2, the degree of drug release achieved a steady state at 6h, with only 9% of MB detected after 2880 min.

The diffusion of MB from the hydrogel network was influenced by the pH of the surrounding medium as MB is a cationic type drug. At low pH, where high level of proton concentration is available, an equilibrium state between MB and protonated water occurred, which reduced the diffusion of MB. Increasing the pH of the buffer solution created osmotic pressure in the surroundings that subsequently increased the diffusion of MB from the hydrogel network through the channels and pores of nanofibers. A low cumulative percentage release of MB in all pHs of the buffer solutions that are less than 22% indicates the potential use of hydrogel derived from sago pulp waste to prolong the MB retaining time and reduce the dosing frequency of MB. This study suggests that the controlled released of MB from hydrogels is also associated with the presence of cross-linkages between the nanofibers polymers, thus restricting the diffusion rate of MB into the surrounding medium.

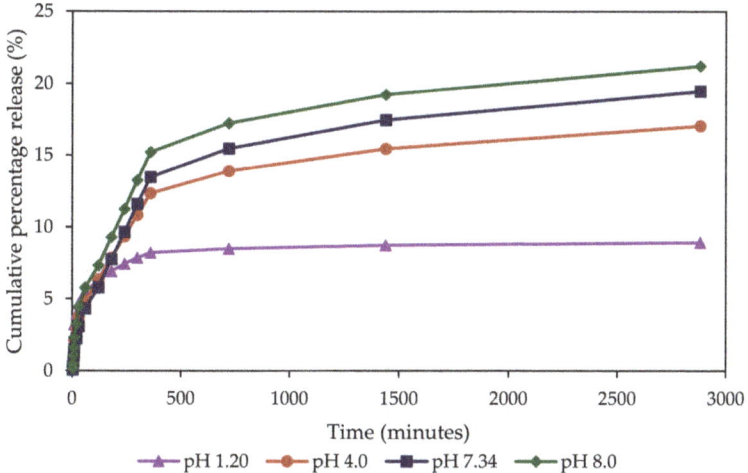

Figure 8. Effects of pH value on the release of MB from CMSP hydrogel nanofibers.

4. Conclusions

This study has investigated the modification of cellulose obtained from sago waste in forming hydrogel nanofibers as a potential drug carrier in wound treatments. The CMSP nanofibers were produced using the electrospinning method with the addition of PEO, which increased the structure and morphology of the nanofibers. The concentration of CMSP/PEO solution, the ratio between CMSP and PEO, and the flow rate of CMSP/PEO during electrospinning, significantly influenced the morphology and size of the nanofibers. An investigation on the percentage of swelling showed that the presence of citric acid and the variation of curing temperatures increased the number of cross-linked networks between the nanofibers to form strong polymeric structures responsible for the swelling process. Drug release studies using MB as a probe molecule showed that hydrogel nanofibers obtained from CMSP has a loading efficiency of 89.20 ± 0.42%. Drug release profiles of MB-loaded hydrogel nanofibers were significantly influenced by the pH of buffer solutions and revealed slow release in all pHs, which perhaps demonstrate the potential use of CMSP hydrogel nanofibers as an in vivo or in vitro drug carrier. This preliminary work has illustrated the potential of CMSP nanofibers as a drug carrier that is not only limited to the incorporation of MB, but also applicable for any of hydrophilic pharmaceutical drugs. Due to the delicate nature of hydrogel nanofibers structure, CMSP/PEO hydrogel nanofibers are more suitable for applications, like wound dressing rather than oral administration.

Author Contributions: The work presented in this paper was a collaboration of all authors. The work contribution among the authors as follow: conceptualization, N.A.R.; methodology, N.M.K. and N.H.R.; software, N.M.K.; validation, N.A.R. and N.M.K., and N.A.R.; formal analysis, N.M.K.; investigation, N.M.K.; resources, N.M.K.; data curation, N.M.K.; writing—original draft preparation, N.M.K.; writing—review and editing, H.B. and H.M.; visualization, N.M.K.; supervision, N.A.R.; project administration, N.A.R.; funding acquisition, N.A.R.

Funding: This research was funded by Universiti Putra Malaysia, grant number GP-IPS/2016/9594900 and Graduate Research Fund (GRF) for Nafeesa Mohd Kanafi scholarship. Check carefully that the details given are accurate and use the standard spelling of funding agency names at https://search.crossref.org/funding, any errors may affect your future funding.

Acknowledgments: Authors would like to acknowledge Universiti Putra Malaysia for financial supports through Research University Grant (Grant code: GP-IPS/2016/9594900). Authors would also acknowledge Graduate Research Fund (GRF) for Nafeesa Mohd Kanafi from Universiti Putra Malaysia.

Conflicts of Interest: The authors declare no conflicts of interest.

References

1. Bujang, K.B.; Hassan, M.A. Recovery of Glucose from Residual Starch of Sago Hampas for Bioethanol Production. *Biomed. Res. Int.* **2013**, *2013*, 1–8.
2. Pushpamalar, V.; Langford, S.J.; Ahmad, M.; Lim, Y.Y. Optimization of reaction conditions for preparing carboxymethyl cellulose from sago waste. *Carbohydr. Polym.* **2006**, *64*, 312–318. [CrossRef]
3. Deparment of Agriculture Sarawak. Available online: https://www.doa.sarawak.gov.my (accessed on 30 July 2018).
4. Veeramachineni, A.K.; Sathasivam, T.; Muniyandy, S. Optimizing Extraction of Cellulose and Synthesizing Pharmaceutical Grade Carboxymethyl Sago Cellulose from Malaysian Sago Pulp. *Appl. Sci.* **2016**, *6*, 170. [CrossRef]
5. Frenot, A.; Henriksson, M.W.; Walkenstro, P. Electrospinning of Cellulose-Based Nanofibers. *J. Appl. Polym. Sci.* **2007**, *103*, 1473–1482. [CrossRef]
6. Kanafi, N.M.; Rahman, N.A.; Rosdi, N.H. Citric acid cross-linking of highly porous carboxymethyl cellulose/poly (ethylene oxide) composite hydrogel films for controlled release applications. *Mater Today Proc.* **2019**, *7*, 721–731. [CrossRef]
7. Shen, X.; Shamshina, J.L.; Berton, P.; Rogers, R.D. Hydrogels based on cellulose and chitin: Fabrication, properties and applications. *Green Chem.* **2015**, *18*, 53–75. [CrossRef]
8. Tan, H.L.; Wong, Y.Y.; Muniyandy, S.; Hashim, K. Carboxymethyl sago pulp/carboxymethyl sago starch hydrogel: Effect of polymer mixing ratio and study of controlled drug release. *J. Appl. Polym. Sci.* **2016**, *43652*, 1–13. [CrossRef]
9. Barbucci, R.; Magnani, A.; Consumi, M. Swelling Behavior of Carboxymethylcellulose Hydrogels in Relation to Cross-Linking pH and Charge Density. *Macromolecules* **2000**, *33*, 7475–7480. [CrossRef]
10. Rizwan, M.; Yahya, R.; Hassan, A.; Yar, M.; Azzahari, A.; Selvanathan, V.; Sonsudin, F.; Abouloula, C. pH sensitive hydrogels in drug delivery: Brief history, properties, swelling, and release mechanism, material selection and applications. *Polymers* **2017**, *9*, 137. [CrossRef]
11. Abrigo, M.; Mcarthur, S.L.; Kingshott, P. Electrospun Nanofibers as Dressings for Chronic Wound Care: Advances, Challenges, and Future Prospects. *Macromol. Biosci.* **2014**, *14*, 772–792. [CrossRef]

12. Tavakoli, J.; Mirzaei, S.; Tang, Y. Cost-effective double-layer hydrogel composites for wound dressing applications. *Polymers* **2018**, *10*, 305. [CrossRef] [PubMed]
13. Raucci, M.G.; Demitri, C.; Giugliano, D.; Benedictis, V.; de Sannino, A.; Ambrosio, L. Effect of citric acid crosslinking cellulose-based hydrogels on osteogenic differentiation. *J. Biomed. Mater. Res.* **2014**, *103A*, 2045–2056. [CrossRef] [PubMed]
14. Arslan, N. Production of carboxymethyl cellulose from sugar beet pulp cellulose and rheological behaviour of carboxymethyl cellulose. *Carbohydr. Polym.* **2003**, *54*, 73–82.
15. Biswas, A.; Kim, S.; Selling, G.W.; Cheng, H.N. Conversion of agricultural residues to carboxymethylcellulose and carboxymethylcellulose acetate. *Ind. Crop. Prod.* **2014**, *60*, 259–265. [CrossRef]
16. Bono, A.; Ying, P.H.; Yan, F.Y.; Muei, C.L.; Sarbatly, R.; Krishnaiah, D. Synthesis and Characterization of Carboxymethyl Cellulose from Palm Kernel Cake. *Adv. Nat. Appl. Sci.* **2009**, *3*, 5–11.
17. Rachtanapun, P.; Luangkamin, S.; Tanprasert, K.; Suriyatem, R. Carboxymethyl cellulose film from durian rind. *YFSTL* **2012**, *48*, 52–58.
18. Pushpamalar, V.; Langford, S.J.; Ahmad, M.; Hashim, K. Preparation of Carboxymethyl Sago Pulp Hydrogel from Sago Waste by Electron Beam Irradiation and Swelling Behavior in Water and Various pH Media. *J. Appl. Polym.* **2013**, *128*, 451–459. [CrossRef]
19. Lyn, Y.; Muniyandy, S.; Kamaruddin, H.; Mansor, A.; Janarthanan, P. Radiation cross-linked carboxymethyl sago pulp hydrogels loaded with ciprofloxacin: Influence of irradiation on gel fraction, entrapped drug and in vitro release. *Radiat. Phys. Chem.* **2015**, *106*, 213–222.
20. Thenapakiam, S.; Kumar, D.G.; Pushpamalar, J.; Saravanan, M. Aluminium and radiation cross-linked carboxymethyl sago pulp beads for colon targeted delivery. *Carbohydr. Polym.* **2013**, *94*, 356–363. [CrossRef]
21. Rathinamoorthy, R.; Technology, F.; College, P.S.G. Nanofiber for drug delivery system. *Pak Text J.* **2012**, *61*, 45–48.
22. Haider, A.; Haider, S.; Kang, I. A comprehensive review summarizing the effect of electrospinning parameters and potential applications of nanofibers in biomedical and biotechnology. *Arab. J. Chem.* **2015**, *11*, 015. [CrossRef]
23. Teck, C. Progress in Polymer Science Nanofiber technology: Current status and emerging developments. *Prog. Polym. Sci.* **2017**, *70*, 1–17.
24. Samadian, H.; Salehi, M.; Farzamfar, S.; Vaez, A.; Ehterami, A.; Sahrapeyma, H.; Goodarzi, A.; Ghorbani, S. In vitro and in vivo evaluation of electrospun cellulose acetate/gelatin/hydroxyapatite nanocomposite mats for wound dressing applications. *Artif. Cells Nanomed. Biotechnol.* **2018**, *46*, 964–974. [CrossRef] [PubMed]
25. Basu, P.; Repanas, A.; Glasmacher, B.; Narendrakumar, U.; Manjubala, I. PEO-CMC blend nanofibers fabrication by electrospinning for soft tissue engineering applications. *Mater. Lett.* **2017**, *195*, 10–13. [CrossRef]
26. Schirmer, R.H.; Coulibaly, B.; Stich, A.; Scheiwein, M.; Merkle, H.; Eubel, J.; Becker, H.; Muller, O.; Zich, T.; Schiek, W. Methylene blue as an antimalarial agent. *Redox Rep.* **2003**, *8*, 272–275. [CrossRef] [PubMed]
27. Oz, M.; Lorke, D.E.; Petroianu, G.A. Methylene blue and Alzheimer's disease. *Biochem. Pharmacol.* **2009**, *78*, 927–932. [CrossRef] [PubMed]
28. Wendel, W.B. The control of methemoglobinemia with methylene blue. *J. Clin. Investig.* **1939**, *18*, 179–185. [CrossRef] [PubMed]
29. Pillay, V.; Dott, C.; Choonara, Y.E.; Tyagi, C.; Tomar, L.; Kumar, P.; Ndesendo, V. MA Review of the Effect of Processing Variables on the Fabrication of Electrospun Nanofibers for Drug Delivery Applications. *J. Nanomat.* **2013**, *2013*, 1–22. [CrossRef]
30. Dahlan, N.A.; Ng, S.L.; Pushpamalar, J. Adsorption of methylene blue onto powdered activated carbon immobilized in a carboxymethyl sago pulp hydrogel. *J. Appl. Polym. Sci.* **2017**, *134*. [CrossRef]
31. Giani, G.; Fredi, S.; Barbucci, R. Hybrid magnetic hydrogel: A potential system for controlled drug delivery by means of alternating magnetic fields. *Polymers* **2012**, *4*, 1157–1169. [CrossRef]
32. Foroutan, H.; Khodabakhsh, M.; Rabbani, M. Investigation of synthesis of PVP hydrogel by irradiation. *Iran J. Radiat. Res.* **2007**, *5*, 131–136.
33. Huang, W.F.; Tsui, G.C.P.; Tang, C.Y.; Yang, M. Optimization Strategy for Encapsulation Efficiency and Size of Drug Loaded Silica Xerogel/Polymer Core-Shell Composite Nanoparticles Prepared by Gelation-Emulsion Method. *Polym. Eng. Sci.* **2017**, *58*, 742–751. [CrossRef]

34. Shi, Y.; Wan, A.; Shi, Y.; Zhang, Y.; Chen, Y. Experimental and Mathematical Studies on the Drug Release Properties of Aspirin Loaded Chitosan Nanoparticles. *BioMed. Res. Int.* **2014**, *2014*. [CrossRef] [PubMed]
35. Chandrasekaran, A.R.; Jia, C.Y.; Theng, C.S.; Muniandy, T.; Muralidharan, S.; Arumugam, S. Invitro studies and evaluation of metformin marketed tablets-Malaysia. *J. Appl. Pharm. Sci.* **2011**, *1*, 214–217.
36. Bolio-López, G.I.; Ross-Alcudia, R.E.; Veleva, L.; Azamar, J.A.; Barrios, G.C.M.; Hernández-Villegas, M.M.; Córdova, S.S. Extraction and Characterization of Cellulose from Agroindustrial Waste of Pineapple (*Ananas comosus* L. Merrill) Crowns. *Chem. Sci. Rev. Lett.* **2016**, *5*, 198–204.
37. Fernandes, J.G.; Correia, D.M.; Botelho, G.; Padrão, J.; Dourado, F. PHB-PEO electrospun fiber membranes containing chlorhexidine for drug delivery applications. *Polym. Test.* **2014**, *34*, 64–71. [CrossRef]
38. Greiner, A.; Wendorff, JH. Electrospinning: A Fascinating Method for the Preparation of Ultrathin Fibers. *Angew. Chem. Int. Ed.* **2007**, *46*, 5670–5703. [CrossRef] [PubMed]
39. Rogina, A. Applied Surface Science Electrospinning process: Versatile preparation method for biodegradable and natural polymers and biocomposite systems applied in tissue engineering and drug delivery. *Appl. Surf. Sci.* **2014**, *296*, 221–230. [CrossRef]
40. Sadat, A.; Hamid, M.; Hajiesmaeilbaigi, F. Effect of electrospinning parameters on morphological properties of PVDF nanofibrous scaffolds. *Prog. Biomater.* **2017**, *6*, 113–123.
41. Megelski, S.; Stephens, J.S.; Chase, D.B.; Rabolt, J.F. Micro and Nanostructured Surface Morphology on Electrospun Polymer Fibers. *Macromolecules* **2002**, *35*, 8456–8466. [CrossRef]
42. Rodoplu, D.; Mutlu, M. Effects of Electrospinning Setup and Process Parameters on Nanofiber Morphology Intended for the Modification of Quartz Crystal Microbalance Surfaces. *J. Eng. Fibers Fabr.* **2012**, *7*, 118–123. [CrossRef]
43. Kumar, S.U.; Matai, I.; Dubey, P.; Bhushan, B.; Sachdev, A. Supporting information Differentially cross-linkable core-shell nanofibers for tunable delivery of anticancer drugs: Synthesis, characterization and its anticancer efficacy. *Electron. Suppl. Mat.* **2014**, *4*, 38263–38272.
44. Khan, S.; Ranjha, N.M. Effect of degree of cross-linking on swelling and on drug release of low viscous chitosan/poly(vinyl alcohol) hydrogels. *Polym. Bull.* **2014**, *71*, 2133–2158. [CrossRef]
45. Reddy, N. Citric acid cross-linking of starch films. *Food Chem.* **2010**, *118*, 702–711. [CrossRef]
46. Tavakoli, J.; Zhang, H.P.; Tang, B.Z.; Tang, Y. Aggregation-induced emission lights up the swelling process: A new technique for swelling characterisation of hydrogels. *Mater. Chem. Front.* **2019**, *3*, 664–667. [CrossRef]
47. Bhattarai, N.; Gunn, J.; Zhang, M. Chitosan-based hydrogels for controlled, localized drug delivery. *Adv. Drug Deliv. Rev.* **2010**, *62*, 83–99. [CrossRef]
48. George, M.; Abraham, T.E. pH sensitive alginate-guar gum hydrogel for the controlled delivery protein drugs. *Int. J. Pharm.* **2007**, *335*, 123–129. [CrossRef]
49. Percival, S.L.; Mccarty, S.; Hunt, J.A.; Woods, E.J. The effects of pH on wound healing, biofilms and antimicrobial efficacy. *Wound Rep. Reg.* **2014**, *22*, 174–186. [CrossRef]

© 2019 by the authors. Licensee MDPI, Basel, Switzerland. This article is an open access article distributed under the terms and conditions of the Creative Commons Attribution (CC BY) license (http://creativecommons.org/licenses/by/4.0/).

Article

Fabrication of Water Absorbing Nanofiber Meshes toward an Efficient Removal of Excess Water from Kidney Failure Patients

Mirei Tsuge [1,2], Kanoko Takahashi [1,2], Rio Kurimoto [2], Ailifeire Fulati [3], Koichiro Uto [2], Akihiko Kikuchi [1] and Mitsuhiro Ebara [1,2,3,*]

1. Graduate School of Industrial Science and Technology, Tokyo University of Science, Katsushika-ku, Tokyo 125-8585, Japan; mirei0927star@gmail.com (M.T.); k.takahashi.lab@gmail.com (K.T.); kikuchia@rs.noda.tus.ac.jp (A.K.)
2. International Center for Materials Nanoarchitectonics (WPI-MANA), National Institute for Materials Science (NIMS), Tsukuba, Ibaraki 305-0044, Japan; kurimoto.rio@gmail.com (R.K.); uto.koichiro@nims.go.jp (K.U.)
3. Graduate School of Pure and Applied Sciences, University of Tsukuba, Tsukuba, Ibaraki 305-8577, Japan; elfirepolat@yahoo.com
* Correspondence: ebara.mitsuhiro@nims.go.jp; Tel.: +81-29-860-4775

Received: 22 February 2019; Accepted: 22 April 2019; Published: 1 May 2019

Abstract: Excellent water-absorbing nanofiber meshes were developed as a potential material for removing excess fluids from the blood of chronic renal failure patients toward a wearable blood purification system without requiring specialized equipment. The nanofiber meshes were successfully fabricated from poly(acrylic acid) (PAA) under various applied voltages by appropriately setting the electrospinning conditions. The electrospun PAA nanofibers were thermally crosslinked via heat treatment and then neutralized from their carboxylic acid form (PAA) to a sodium carboxylate form poly(sodium acrylate) (PSA). The PSA nanofiber meshes exhibited a specific surface area 393 times that of the PSA film. The PSA fiber meshes showed a much faster and higher swelling than its corresponding film, owing to the higher capillary forces from the fibers in addition to the water absorption of the PSA gel itself. The proposed PSA fibers have the potential to be utilized in a new approach to remove excess water from the bloodstream without requiring specialized equipment.

Keywords: water absorbing materials; nanofibers; electrospinning; poly(sodium acrylate); hemodialysis

1. Introduction

The main function of the kidneys is to filter blood to remove waste products such as uremic toxins and excess water in the form of urine. Kidney failure or renal failure is the last stage of chronic kidney disease in which the kidneys no longer function, causing impaired consciousness, heart failure, pulmonary edema, and subcutaneous bleeding [1]. The most common treatment for kidney failure is hemodialysis (HD). As of 2015, it is reported that approximately 2.6 million patients undergo regular dialysis treatments worldwide, and the number of patients is expected to increase to 5.6 million by 2030 [1–3].

One of the disadvantages of HD treatment is that it requires substantial resources, including 150+ liters of water, because the mechanism of HD is based on a diffusion-limited process wherein the blood flows along an external semipermeable membrane and excess wastes or fluids pass into a dialysate solution [4,5]. Despite recent improvements in the efficiency and selectivity properties of dialysis membranes, the diffusion-based HD process is inconvenient, time-consuming, and expensive [6]. In particular, HD treatments have been severely restricted in developing countries and during natural disasters because of the lack of dialysis supplies and frequent electricity outages, resulting in over

1 million people worldwide die from untreated kidney failure annually [7–10]. Notably, during the 2011 Tohoku earthquake and tsunami in Japan, many dialysis systems were damaged, and large shortages of dialysate, water, and electricity created an extremely dangerous situation for HD-dependent patients [11,12]. Therefore, it is necessary to develop simpler and more-accessible methods for kidney failure treatments worldwide. From these perspectives, we focused on utilizing selective absorption as a technique for blood purification. In particular, we focused on a water-absorbing mechanism (materials) instead of a diffusion-based process to avoid the use of 150+ liters of water. In our previous work, we prepared a zeolite–polymer composite nanofiber mesh to selectively absorb the uremic toxin creatinine using a biocompatible polymer, namely poly(ethylene-co-vinyl alcohol) [13,14]. Nanofiber meshes have many advantages, such as large specific surface areas, excellent mechanical properties, low pressure losses, and ease of handling [15,16]. Several techniques are available for synthesizing nanofibers. Among them, electrospinning method has been demonstrated to provide promising results in various fields [17–21]. Electrospinning is a relatively inexpensive and simple method for producing a variety of nano- and micro-scale fibers with uniform diameters [22,23].

In addition to creatinine, uremic toxins (such as urea), as well as excess water and electrolytes should be removed through dialysis treatments. A high water content in the body is directly linked to weight gain, swelling, breathing difficulty, hypertension, heart failure, and pulmonary edema [7–10]. In this work, we focus on poly(sodium acrylate) (PSA) gels as a potential material for removing excess fluids from the blood of chronic renal failure patients. PSA is a strongly hydrophilic polymer owing to its pendent carboxylate anions, and it can absorb more than 10 times its own weight to form a gel. Therefore, it is commonly used in disposable diapers and sanitary supplies [24–26]. The water absorption ability can be improved by preparing nanofiber meshes out of PSA owing to the higher capillary forces from the fibers in addition to the water absorption of the PSA gel itself. In this study, we developed a PSA nanofiber mesh as a wearable blood purification system without requiring specialized equipment.

2. Materials and Methods

2.1. Materials

Poly(acrylic acid) (PAA) (average M_V = 250,000 g mol^{-1}), ethylene glycol (EG) (anhydrous), sodium citrate, sulfuric acid, ethanol, and methanol were purchased from Wako Pure Chemicals (Osaka, Japan). Poly(vinyl pyrrolidone) (PVP; M_W = 360,000 g mol^{-1}) and Dulbecco's phosphate buffered saline were purchased from Sigma-Aldrich (Madison, WI, USA) and used without further purification. Female severe combined immunodeficiency (SCID) mice 6 weeks of age were obtained from Charles River Laboratories Japan, Inc. (Yokohama, Japan). All animal care and experimental procedures were approved by the Experimental Animal Administration Committee of National Institute for Materials Science (Approval No. 36-2013-2).

2.2. Fiber Fabrication

Electrospinning solutions were prepared by dissolving 8 or 10 wt/v% PAA in methanol or ethanol. EG was added to each sample as a crosslinking agent at concentrations of 10 and 16 wt% relative to the PAA, and complete dissolution was observed after 24 h of mixing using a magnetic stirrer at room temperature. Sulfuric acid (1 mol L^{-1}) was added to the PAA–EG solution immediately before electrospinning at a concentration of 50 μL mL^{-1}. Additionally, PVP ethanol solutions were prepared, with a concentration of 10 wt/v%. The electrospinning was performed using a NANON-03A electrospinning machine (MECC, Fukuoka, Japan). To facilitate the removal of electrospun PAA from the aluminum foil, a thin layer of PVP fibers were deposited by electrospinning the PVP solutions prior to PAA as a sacrificial layer. For the PVP fibers, the following parameters were kept constant: an applied voltage of 15 kV, a flow rate of 0.8 mL h^{-1}, and a working distance of 15 cm. A 25-gauge pointed needle was used. For the PAA fibers, the spinning parameters were kept constant: a flow rate of 0.6 mL h^{-1} and a working distance of 10 cm; a 25-gauge pointed needle was used, and the applied voltages selected were 10, 15, 20, 25, and 30 kV. The fibers were electrospun onto a aluminum foil sheet placed on a stationary plate collector. The electrospun PAA nanofibers were thermally crosslinked on the mandrel via heat treatment in a vacuum oven (at 130 °C with a reduced pressure of 25 mmHg (84.7 kPa)) for 30 min and then cooled to room temperature. Prior to characterization, the electrospun fibers were removed from the foil in water to dissolve and wash off the sacrificial layer. The PAA nanofibers were neutralized from their carboxylic acid form (PAA) to a sodium carboxylate form (PSA) by immersing them in a 1 mol L^{-1} NaOH and 1 mol L^{-1} NaCl solution for approximately 1 h, then rinsing with water to remove any residual salts. To confirm the success of intermolecular crosslinking after the reaction, the disappearance of the carboxy groups in PAA was observed by ATR–FTIR spectroscopy (Thermoscientific Nicolet 4700, Thermo, Waltham, MA, USA). Figure 1 shows the fabrication process of the PSA nanofiber mesh using the electrospinning method.

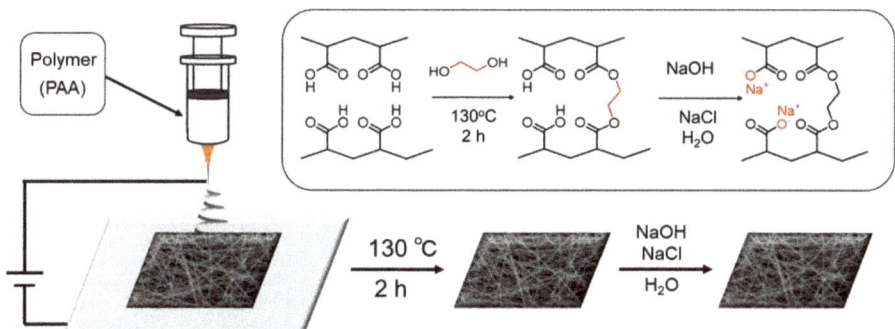

Figure 1. Schematic illustration of the fabrication process of a poly(sodium acrylate) (PSA) nanofiber mesh using the electrospinning method.

2.3. Fiber Characterization

The fiber morphologies were observed using a Hitachi S-4800 field emission scanning electron microscope (FESEM) (Hitachi, Tokyo, Japan). Prior to imaging, the samples were sputter coated with ~3 nm platinum using an E-1030 Ion Sputterer (15 mA, 30 s, 6.0 Pa) (Hitachi). Lower-resolution images were obtained using a NeoScope JCM-5000 benchtop SEM (JEOL, Tokyo, Japan). The fiber diameters were measured using the ImageJ software (Image J 1.52a USAJAVA 1.8.0_112 (64-bit), the National Institutes of Health, Bethesda, MD, USA) and presented as an average with standard deviations (n = 40). The N$_2$ gas adsorption/desorption isotherms and Brunauer–Emmett–Teller (BET) surface area were characterized using a BELSORP-mini ASAP 2020 analyzer (BEL Japan, Inc., Osaka, Japan).

2.4. Swelling Behavior of Nanofiber Meshes and Films

The swelling ratio of the nanofiber mesh and film were determined from the relationship

$$\text{swelling ratio} = (W_S - W_0)/W_0, \tag{1}$$

where W_0 is the dry weight of the PAA (or PSA), and W_S is the weight of the swollen sample immersed in the solution at room temperature, after the excess water on the sample surfaces was imbibed with a Kimwipe (n = 3).

3. Results and Discussion

3.1. Fabrication of PSA Nanofiber Meshes

In this work, nanofiber meshes made of a water-absorbing polymer (i.e., PSA), with a hydrophilic sodium carboxylate group were prepared. First, PVP nanofibers were formed as a sacrificial layer by electrospinning, followed by electrospinning the PAA nanofibers from solutions of varying concentrations, solvents, and voltages. The PAA nanofiber mesh was neutralized to its PSA form and examined using the FESEM. As shown in Figure 2, stable PAA nanofiber meshes were observed regardless of solvents (ethanol and methanol). When preparing nanofiber meshes at different concentrations, we found that 10 wt/v% PAA produced homogeneous fibers with less dispersity compared with the 8 wt/v% solutions. We could not fabricate fine nanofibers at a concentration of below 6 or above 10 wt/v%. The effect of voltage on the PAA fiber diameter was also examined for those made with 10 wt/v% solutions in methanol (Figure 2). The combined action of the Coulombic forces, surface tension, and viscosity tends to affect the diameter of the electrospun nanofibers (Figure 3). The nanofibers electrospun at low potentials tend to have large diameters with the presence of beads because of the weaker Coulombic forces relative to the surface tension and viscoelasticity. With an increase in the applied voltage, the Coulombic forces increased, and fibers with a narrow diameter distribution were formed when the three forces balanced each other. The voltage varying in the range of 10–30 kV applied to the electrospun PAA is consistent with these principles, as the fiber diameters decreased with the increase in the voltage. Based on these experimental results, we found that PAA fibers prepared in 10 wt/v% of methanol at 30 kV were optimal, owing to their high surface area and narrow fibers.

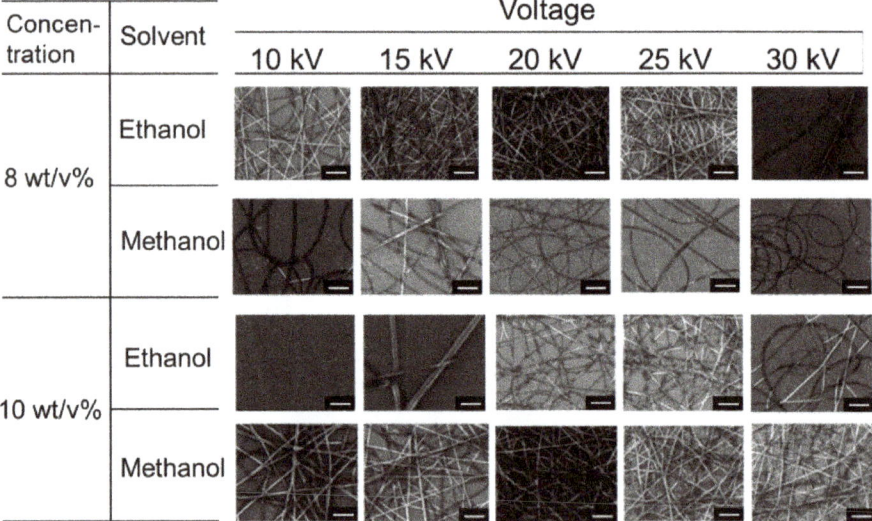

Figure 2. Effects of applied voltage and polymer concentration on the poly(acrylic acid) (PAA) fiber formation and morphologies (scale bar: 10 μm).

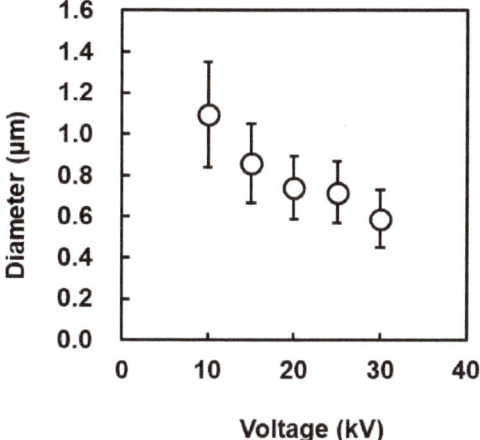

Figure 3. Effect of applied voltage on the PAA fiber diameter.

3.2. Porous Structure Analysis of PSA Nanofiber Meshes

To analyze the specific surface area of the nanofiber meshes and the structure between the fibers, a nitrogen adsorption test was carried out on vacuum-dried PSA nanofiber meshes. The adsorption/desorption isotherms (Figure 4) show typical IV-type properties with an H3 hysteresis loop. From this hysteresis loop, we can conclude that the nanofiber meshes have a mesoporous structure [27–29]. The calculated BET surface area was 1.6536 m^2/g, while that of the corresponding film was 0.004 m^2/g. Therefore, the nanofiber mesh was found to have a specific surface area 393 times that of the PSA films. Stable fibers were formed when electrospinning PAA meshes with EG as a crosslinking agent, regardless of the EG concentration (Figure 5). In addition, the concentration of the crosslinking agent had little effect on the fiber diameter. Figure 5 shows the SEM images of the PAA nanofiber meshes added with 0, 0.16, 1.6, and 16 wt% of the EG crosslinking agent.

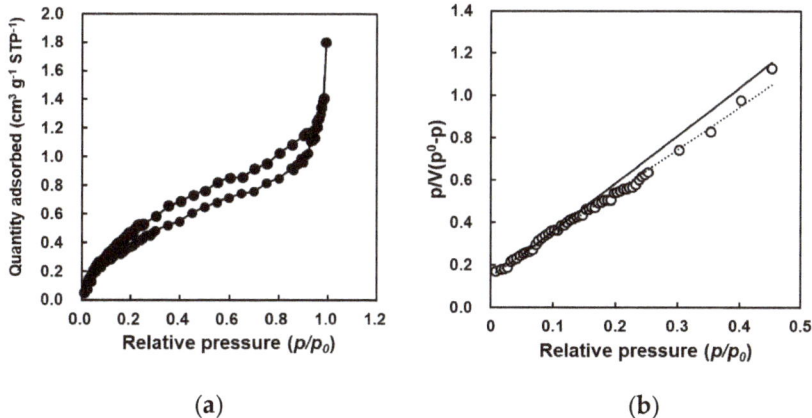

Figure 4. (a) N$_2$ adsorption–desorption isotherms of PSA nanofiber meshes; (b) Brunauer–Emmett–Teller (BET) plot of PSA nanofiber meshes.

Next, the stability of the fibers with different crosslinking densities was investigated by immersing them in water. In Figure 6a, no fibers are observed because the PSA dissolved into the solution.

Increasing the EG concentration to 0.16% resulted in an ill-defined structure with no stable fibers (Figure 6b). A fibrous network was obtained at an EG concentration of 1.6 wt%; however, the individual fibers were highly swollen and adhered to each other. At 16 wt% EG crosslinking, a stable nanofiber mesh with well-defined PSA fibers was observed. The success of intermolecular crosslinking was also confirmed by ATR–FTIR spectroscopy (Supplementary Materials: Figure S6). These results demonstrate that the structure of the PSA nanofiber mesh can be changed by varying the amount of EG, with an improved fiber stability at higher crosslinking densities.

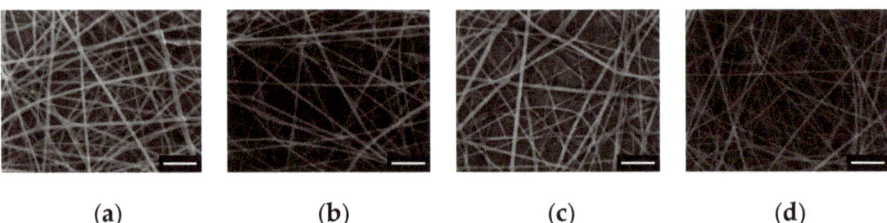

Figure 5. SEM images of PAA nanofibers in a dry state. The nanofibers were prepared by adding ethylene glycol at concentrations of (**a**) 0 wt%; (**b**) 0.16 wt%; (**c**) 1.6 wt%; (**d**) 16 wt% (scale bar: 10 μm).

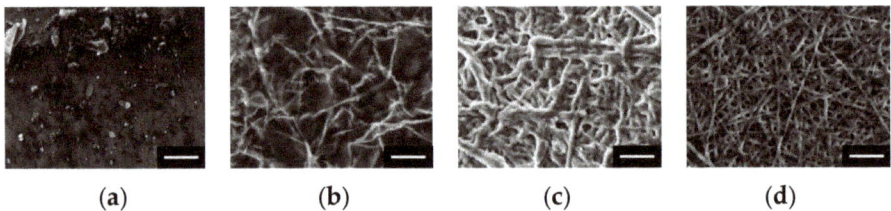

Figure 6. SEM images of PSA nanofibers after immersing in water. The nanofibers were prepared by adding ethylene glycol at concentrations of (**a**) 0 wt%; (**b**) 0.16 wt%; (**c**) 1.6 wt%; (**d**) 16 wt% (scale bar: 10 μm).

3.3. Water Absorption Test on PSA Nanofiber Meshes

To test the water absorption properties of our nanofibers, dried PSA meshes or films prepared with various concentrations of the EG crosslinking agent were immersed in phosphate buffered saline (PBS), and the masses of the fibers or films were measured every 5 min until equilibrium absorption was reached (Figure 7). As shown in Figure 7a,b, the swelling ratio of both the nanofiber meshes and films decreased with an increase in the concentration of the crosslinking agent. This trend suggests an increase in the crosslinking density of the PSA fibers and films at higher EG concentrations. Comparing the swelling ratios of the nanofiber meshes and films at each concentration, we find that the final swelling ratios were very similar under most conditions. However, a remarkable difference is observed in the swelling rate at an EG concentration of 0.16 wt%. Specifically, the swelling ratio of the nanofiber meshes quickly reached its maximum in 5 min, whereas the film took 40 min to reach equilibrium swelling. This difference is attributed to the high surface area of the PSA nanofiber mesh relative to the film. However, at an EG concentration of 1.6 wt%, there was no significant difference in the swelling ratio or swelling rate between the meshes and the films. In this case, the difference in the surface area did not lead to any noticeable changes in the swelling speeds, probably because the concentration of the crosslinking agent and the crosslinking density were too high. At an EG concentration of 16 wt%, little to no swelling was observed in the PSA film, whereas significant swelling occurred in the mesh. This difference in the swelling ratio can be attributed to the high capillary forces of the stable nanofibers at higher crosslinking densities. Swelling due to the capillary forces of the fibers was not observed at EG concentrations of 0.16 and 1.6 wt% because of their collapsed structure, as shown in Figure 6.

Subsequently, a water absorption test was carried out on the PSA nanofiber meshes and films prepared with an EG concentration of 16 wt% in mouse blood. The PSA nanofiber mesh exhibited a swelling ratio approximately 10 times its own weight, whereas the swelling of the PSA film was 2.5 times lower at an approximate swelling ratio of 4 (Figure 8). This result is consistent with the previous absorption test conducted in PBS, wherein the nanofiber mesh exhibited a higher swelling ratio because of its increased surface area and capillary forces. The SEM images of the PSA nanofibers taken after the water absorption test in the mouse blood show that the meshes retained their shape and that the fiber diameter remained the same. However, some components of the blood were found to adhere to the fiber surfaces. In comparison, although the film was flat before the water absorption test, the blood components adhered to the entire surface of the film after the water absorption test. Based on this, we can conclude that blood components are less likely to adhere to the fiber than to the film. However, PSA itself is not excellent biocompatible material and, therefore, use of anti-coagulant is still needed when we apply our material to the blood in the future.

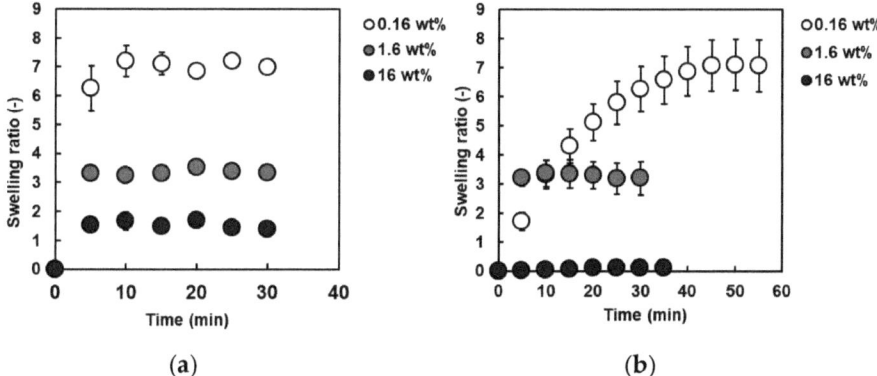

Figure 7. Water absorption of PSA nanofiber meshes and films prepared with different crosslinker concentrations in PBS (pH: 7.4). Ethylene glycol concentration: closed circle; 16 wt%; shaded circle, 1.6 wt%; and open circle, 0.16 wt%; respectively. Data are expressed as mean ± standard deviation (SD) (n = 40). Absorption of PAA nanofiber mesh/film in phosphate buffered saline (PBS): (a) fiber; (b) film.

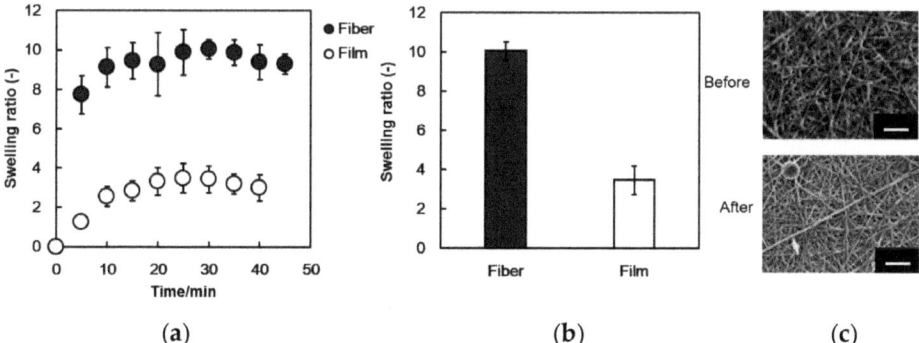

Figure 8. (a,b) Water absorption of PSA nanofiber meshes (closed circle) and PSA film (open circle) in mouse blood. Data are expressed in mean ± SD (n = 3); (c) SEM images of the PSA nanofibers before (upper images) and after (lower images) the water absorption test in blood. Scale bar: 10 μm.

4. Conclusions

This study investigated the water absorption ability of PSA fibers prepared by the electrospinning process. The fiber meshes were successfully fabricated from PAA under various applied voltages by appropriately setting the electrospinning conditions. The applied voltage was a key parameter, the optimum value of which was found to be 30 kV. The PSA nanofiber meshes exhibited a specific surface area 393 times that of the PSA film. The fabricated PSA fibers could absorb water from solution and blood. Although other uremic toxins need to be removed, the proposed water absorbing fibers have the potential to be utilized as a new filter in a wearable blood purification system, particularly in disaster-hit sites and the developing world.

Supplementary Materials: The following are available online at http://www.mdpi.com/2079-6439/7/5/39/s1, Figure S1: SEM images of the PVP nanofibers. (a) 10 kV; (b) 15 kV; (c) 20 kV, Figure S2: PVP fiber diameter dependence on electrospinning voltage, Figure S3: Histogram of PAA nanofiber diameters, Figure S4: SEM images of the PAA nanofibers (a) without (b) with heat treatment, Figure S5: Effect of concentration of crosslinking agent on the fiber diameter, Figure S6: ATR-FTIR spectra of PSA nanofiber mesh before and after crosslinking reaction.

Author Contributions: Conceptualization, M.E.; Methodology, M.T., R.K., and K.T.; Software, M.T.; Formal Analysis, M.T., K.U., A.F.; Investigation, M.T.; Resources, M.E.; Data Curation, M.T. and K.T.; Writing—Original Draft Preparation, M.T.; Writing—Review & Editing, M.E. and A.K.; Supervision, M.E.; Project Administration, M.E.; Funding Acquisition, M.E.

Funding: This research was funded by JSPS KAKENHI (Grant Number 264086 and 26750152).

Acknowledgments: The authors thank the support from Namiki Foundry at National Institute for Materials Science (NIMS) for SEM measurements.

Conflicts of Interest: The authors declare no conflict of interest.

References

1. Couser, W.G.; Remuzzi, G.; Mendis, S.; Tonelli, M. The contribution of chronic kidney disease to the global burden of major noncommunicable diseases. *Kidney Int.* **2011**, *80*, 1258–1270. [CrossRef] [PubMed]
2. Liyanage, T.; Ninomiya, T.; Jha, V.; Neal, B.; Patrice, H.M.; Okpechi, I.; Zhao, M.; Lv, J.; Garg, A.X.; Knight, J.; et al. Worldwide access to treatment for end-stage kidney disease: A systematic review. *Lancet* **2015**, *385*, 1975–1982. [CrossRef]
3. Glassock, R.J.; Winearls, C. The global burden of chronic kidney disease: How valid are the estimates? *Nephron Clin. Pract.* **2008**, *110*, 39–47. [CrossRef] [PubMed]
4. Harris, A.; Cooper, B.A.; Li, J.J.; Bulfone, L.; Branley, P.; Collins, J.F.; Craig, J.C.; Fraenkel, M.B.; Johnson, D.W.; Kesselhut, J.; et al. Cost-effectiveness of initiating dialysis early: A randomized controlled trial. *Am. J. Kidney Dis.* **2011**, *57*, 707–715. [CrossRef]
5. Baboolal, K.; McEwan, P.; Sondhi, S.; Spiewanowski, P.; Wechowski, J.; Wilson, K. The cost of renal dialysis in a UK setting—A multicentre study. *Nephrol. Dial. Transplant.* **2008**, *23*, 1982–1989. [CrossRef]
6. Young, B.A.; Chan, C.; Blagg, C.; Lockridge, R.; Golper, T.; Finkelstein, F.; Shaffer, R.; Mehrotra, R. How to overcome barriers and establish a successful home HD program. *Clin. J. Am. Soc. Nephrol.* **2012**, *7*, 2023–2032. [CrossRef]
7. Johnson, DW.; Hayes, B.; Gray, NA.; Hawley, C.; Hole, J.; Mantha, M. Renal services disaster planning: Lessons learnt from the 2011 Queensland floods and North Queensland cyclone experiences. *Nephrology* **2013**, *18*, 41–46. [CrossRef] [PubMed]
8. See, E.J.; Alrukhaimi, M.; Ashuntantang, G.E.; Bello, A.K.; Bellorin-Font, E.; Gharbi, M.B.; Braam, B.; Feehally, J.; Harris, D.C.; Jha, V.; et al. Global coverage of health information systems for kidney disease: Availability, challenges, and opportunities for development. *Kidney Int. Suppl.* **2018**, *8*, 74–81. [CrossRef] [PubMed]
9. Murakami, N.; Siktel, H.B.; Lucido, D.; Winchester, J.F.; Harbord, N.B. Disaster preparedness and awareness of patients on hemodialysis after Hurricane Sandy. *Clin. J. Am. Soc. Nephrol.* **2015**, *10*, 1389–1396. [CrossRef] [PubMed]

10. Gorham, G.; Howard, K.; Togni, S.; Lawton, P.; Hughes, J.; Majoni, S.W.; Brown, S.; Barnes, S.; Cass, A. Economic and quality of care evaluation of dialysis service models in remote Australia: Protocol for a mixed methods study. *BMC Health Serv. Res.* **2017**, *17*, 320. [CrossRef] [PubMed]
11. Sever, M.S.; Erek, E.; Vanholder, R.; Kalkan, A.; Guney, N.; Usta, N.; Yilmaz, C.; Kutanis, C.; Turgut, R.; Lameire, N. Features of chronic hemodialysis practice after the Marmara Earthquake. *J. Am. Soc. Nephrol.* **2004**, *15*, 1071–1076. [CrossRef] [PubMed]
12. Nangaku, M.; Akizawa, T. Diary of a Japanese nephrologist during the present disaster. *Kidney Int.* **2011**, *79*, 1037–1039. [CrossRef] [PubMed]
13. Namekawa, K.; Schreiber, M.T.; Aoyagi, T.; Ebara, M. Fabrication of zeolite–polymer composite nanofibers for removal of uremic toxins from kidney failure patients. *Biomater. Sci.* **2014**, *2*, 674–679. [CrossRef]
14. Takai, R.; Kurimoto, R.; Nakagawa, Y.; Kotsuchibashi, Y.; Namekawa, K.; Ebara, M. Towards a rational design of zeolite-polymer composite nanofibers for efficient adsorption of creatinine. *J. Nanomater.* **2016**, *2016*, 1–7. [CrossRef]
15. Vasita, R.; Katti, D.S. Nanofibers and their applications in tissue engineering. *Int. J. Nanomed.* **2006**, *1*, 15–30. [CrossRef]
16. Zhou, F.-L.; Gong, R.-H. Manufacturing technologies of polymeric nanofibres and nanofibre yarns. *Polym. Int.* **2008**, *57*, 837–845. [CrossRef]
17. Greiner, A.; Wendorff, J.H. Electrospinning: A fascinating method for the preparation of ultrathin fibers. *Angew. Chem. Int. Ed.* **2007**, *46*, 5670–5703. [CrossRef]
18. Doshi, J.; Reneker, D.H. Electrospinning process and applications of electrospun fibers. *J. Electrost.* **1995**, *35*, 151–160. [CrossRef]
19. Reneker, D.H.; Yarin, A.L.; Fong, H.; Koombhongse, S. Bending instability of electrically charged liquid jets of polymer solutions in electrospinning. *J. Appl. Phys.* **2000**, *87*, 4531–4547. [CrossRef]
20. Thompson, C.J.; Chase, G.G.; Yarin, A.L.; Reneker, D.H. Effects of parameters on nanofiber diameter determined from electrospinning model. *Polymer* **2007**, *48*, 6913–6922. [CrossRef]
21. Rajala, J.W.; Shin, H.U.; Lolla, D.; Chase, G.G. Core–Shell electrospun hollow aluminum oxide ceramic fibers. *Fibers* **2015**, *3*, 450–462. [CrossRef]
22. Bou, S.; Ellis, A.V.; Ebara, M. Synthetic stimuli-responsive 'smart' fibers. *Curr. Opin. Biotechnol.* **2016**, *39*, 113–119. [CrossRef] [PubMed]
23. Garrett, R.; Niiyama, E.; Kotsuchibashi, Y.; Uto, K.; Ebara, M. Biodegradable nanofiber for delivery of immunomodulating agent in the treatment of basal cell carcinoma. *Fibers* **2015**, *3*, 478–490. [CrossRef]
24. Li, A.; Zhang, J.; Wang, A. Synthesis, characterization and water absorbency properties of poly(acrylic acid)/sodium humate superabsorbent composite. *Polym. Adv. Technol.* **2005**, *16*, 675–680. [CrossRef]
25. Chen, Y.; Tan, H. Crosslinked carboxymethylchitosan-g-poly(acrylic acid) copolymer as a novel superabsorbent polymer. *Carbohydr. Res.* **2006**, *341*, 887–896. [CrossRef] [PubMed]
26. Omidian, H.; Hashemi, S.A.; Sammes, P.G.; Meldrum, I. Modified acrylic-based superabsorbent polymers (dependence on particle size and salinity). *Polymer* **1999**, *40*, 1753–1761. [CrossRef]
27. Fu, Q.; Wang, X.; Si, Y.; Liu, L.; Yu, J.; Ding, B. Scalable fabrication of electrospun nanofibrous membranes functionalized with citric acid for high-performance protein adsorption. *Appl. Mater. Interfaces* **2016**, *8*, 11819–11829. [CrossRef]
28. Si, Y.; Wang, X.; Li, Y.; Chen, K.; Wang, J.; Yu, J.; Wang, H.; Ding, B. Optimized colorimetric sensor strip for mercury(II) assay using hierarchical nanostructured conjugated polymers. *J. Mater. Chem. A* **2014**, *2*, 645–652. [CrossRef]
29. Barakat, N.A.M.; Khalil, K.A.; Mahmoud, I.H.; Kanjwal, M.A.; Sheikh, F.A.; Kim, H.Y. CoNi bimetallic nanofibers by electrospinning: Nickel-based soft magnetic material with improved magnetic properties. *J. Phys. Chem. C* **2010**, *114*, 15589–15593. [CrossRef]

© 2019 by the authors. Licensee MDPI, Basel, Switzerland. This article is an open access article distributed under the terms and conditions of the Creative Commons Attribution (CC BY) license (http://creativecommons.org/licenses/by/4.0/).

Communication

Increased Mechanical Properties of Carbon Nanofiber Mats for Possible Medical Applications

Marah Trabelsi [1,2], Al Mamun [1], Michaela Klöcker [1], Lilia Sabantina [1], Christina Großerhode [1], Tomasz Blachowicz [3] and Andrea Ehrmann [1,*]

[1] Faculty of Engineering and Mathematics, Bielefeld University of Applied Sciences, ITES, 33619 Bielefeld, Germany; marah.trabelsi@enis.tn (M.T.); al.mamun@fh-bielefeld.de (A.M.); michaela.kloecker@fh-bielefeld.de (M.K.); lilia.sabantina@fh-bielefeld.de (L.S.); christina.grosserhode@fh-bielefeld.de (C.G.)
[2] Ecole Nationale d'Ingénieurs de Sfax, Sfax 3038, Tunisia
[3] Institute of Physics-Center for Science and Education, Silesian University of Technology, 44-100 Gliwice, Poland; tomasz.blachowicz@polsl.pl
* Correspondence: andrea.ehrmann@fh-bielefeld.de

Received: 30 September 2019; Accepted: 13 November 2019; Published: 17 November 2019

Abstract: Carbon fibers belong to the materials of high interest in medical application due to their good mechanical properties and because they are chemically inert at room temperature. Carbon nanofiber mats, which can be produced by electrospinning diverse precursor polymers, followed by thermal stabilization and carbonization, are under investigation as possible substrates for cell growth, especially for possible 3D cell growth applications in tissue engineering. However, such carbon nanofiber mats may be too brittle to serve as a reliable substrate. Here we report on a simple method of creating highly robust carbon nanofiber mats by using electrospun polyacrylonitrile/ZnO nanofiber mats as substrates. We show that the ZnO-blended polyacrylonitrile (PAN) nanofiber mats have significantly increased fiber diameters, resulting in enhanced mechanical properties and thus supporting tissue engineering applications.

Keywords: electrospinning; nanofiber mat; ZnO; polyacrylonitrile (PAN); brittleness

1. Introduction

Electrospinning allows for the creation of thin fibers in the diameter range between some ten and several hundred micrometers from diverse polymers or polymer blends [1]. Such nanofibers can be used, for example, for air filtration [2–4], water filtration [5–7], batteries [8] or biomedical applications [9–11].

One of the typical materials often used as a precursor of carbon nanofibers is polyacrylonitrile (PAN) [12,13]. This polymer has the additional advantage of being spinnable from the low-toxic solvent dimethyl sulfoxide (DMSO) [14].

Such PAN nanofiber mats can afterwards be stabilized, typically in air, to enable a chemical cyclization process, including the oxidation, aromatization, dehydrogenation, crosslinking and formation of a thermally stable aromatic ladder polymer [15–18]. This first thermal treatment can be performed, for example, by heating the nanofiber mat to temperatures around 260 °C–290 °C, approached with heating rates between 0.5 K/min and 5 K/min [19–21], but can also be performed with higher temperatures [22] or split into two different stages [23].

A problem which is scarcely discussed in the scientific literature, however, is the dimensional change of the nanofiber mats during stabilization and carbonization, typically linked to a morphological change of the nanofibers themselves, which become thicker, shorter and wavier due to the relaxation of the frozen strain, resulting from the severe stretching during electrospinning [21]. Different strategies are discussed in the literature to avoid this morphological change, which is also connected to a loss of the mechanical stability of the carbon nanofiber mats. Typically, bundles of nanofibers are stretched during stabilization [24–26] and partly even during carbonization [27]. However, this process cannot be transferred to nanofiber mats created by needleless electrospinning. The latter can be fixed instead, which often leads to shrinking in one direction, if only two sides are fixed, or breaking of the mat even at low heating rates [28–30]. To enable the fixing of the nanofiber mats, not only along the borders, but over the whole plane, the nanofiber mats can be electrospun on aluminum foils as substrates and stabilized on these substrates, from which they are separated afterwards [31]. However, this method is better suited for the batch production of nanofiber mats than for continuous production, which is typically performed on a textile substrate delivered to the spinning chamber by a roll-to-roll process.

Here we report on another method to unambiguously retain the original nanofiber morphology during carbonization and, most importantly, to create soft, bendable carbon nanofiber mats. By adding ZnO to the PAN solution, significantly thicker nanofibers are created during electrospinning, resulting in an unchanged morphology during carbonization. ZnO can be used as a gas sensor [32] or for photocatalytic degradation [33]. It is known to be non-cytotoxic, but shows high antibacterial activity against *Staphylococcus aureus* and *Escherichia. coli* [34], making such fibers well suited for medical applications such as wound dressings [35] or as a substrate for cell growth in tissue engineering [36].

2. Materials and Methods

The needleless electrospinning machine "Nanospider Lab" (Elmarco Ltd., Liberec, Czech Republic) was used to prepare nanofiber mats. We applied the following spinning parameters: high voltage 80 kV, nozzle diameter 0.9 mm, carriage speed 100 mm/s, distance between bottom electrode and substrate 240 mm, distance between ground electrode and substrate 50 mm, temperature in the spinning chamber 24 °C and relative humidity 33%. The spinning duration was 20 min (areal weight 4.7 g/m^2).

The spinning solution used here was prepared by dissolving 15 wt.% PAN and 5 wt.% ZnO (nano powder < 100 nm particle size, Sigma-Aldrich, Germany) in DMSO (min. 99.9%, purchased from S3 Chemicals, Bad Oeynhausen, Germany). For comparison, pure PAN nanofiber mats were electrospun from 16 wt.% PAN dissolved in DMSO. The slightly higher amount of PAN was necessary to reach a sufficient viscosity of the polymer solution and thus to avoid the formation of beads along the fibers. Especially for dyeing tests, we used a woven cotton fabric with thickness 0.19 mm and areal weight 84.49 g/m^2.

A muffle furnace B150 (Nabertherm, Lilienthal, Germany) was used to stabilize the nanofiber mats, using a heating rate of 1 K/min to reach a maximum temperature of 280 °C, followed by isothermal treatment for 1 h.

For carbonization, a furnace CTF 12/TZF 12 (Carbolite Gero, Neuhausen, Germany) was used, in which a temperature of 500 °C was reached with a heating rate of 10 K/min, applying a nitrogen flow of 150 mL/min (standard temperature and pressure, STP), again followed by isothermal treatment for 1 h.

For dyeing tests, a forest fruit tea solution (Mayfair, Wilken Tee GmbH, Fulda, Germany) was applied. For this, 2.5 g tea was mixed with 30 g distilled water, and the tea was extracted for 30 min. Samples were dyed for 30 min in a solution that contained anthocyanins as dye molecules and were dried at room temperature.

Investigations of the sample surfaces were performed using a digital microscope VHX-600D (Keyence, Neu-Isenburg, Germany), as well as a scanning electron microscope (SEM) Zeiss 1450VPSE (Oberkochen, Germany) for more detailed examinations. Chemical evaluation was performed by Fourier-transform infrared (FTIR) spectroscopy, using an Excalibur 3100 (Varian, Inc., Palo Alto, CA, USA). Fiber diameters were investigated using ImageJ 1.51j8 (from National Institutes of Health, Bethesda, MD, USA).

3. Results and Discussion

The morphology of the raw electrospun PAN/ZnO nanofiber mats is depicted in Figure 1, showing SEM images with identical magnification, taken at different sample positions. Generally, the fibers are much thicker than common PAN nanofibers, which have average diameters of 160 ± 40 nm, electrospun under identical conditions [31], or the PAN nanofibers created here, resulting in an average diameter of 140 ± 40 nm, as calculated from SEM images. For the PAN/ZnO nanofibers under investigation here, the average diameter is 450 ± 120 nm, i.e., approximately three times the common PAN nanofiber diameter.

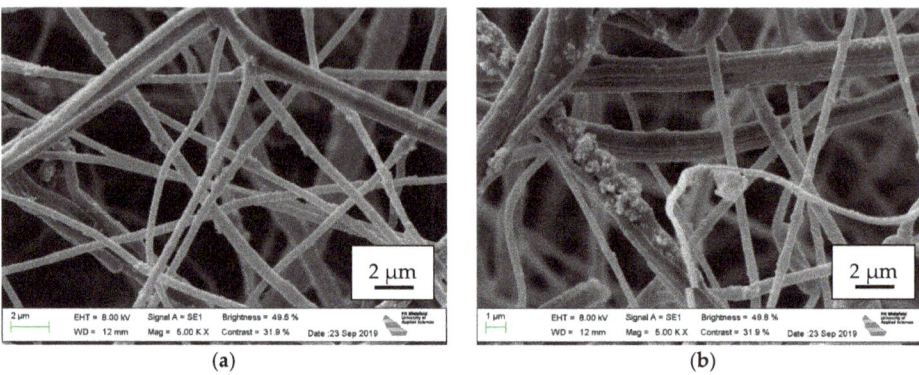

Figure 1. Scanning electron microscopy SEM images of an electrospun polyacrylonitrile (PAN)/ZnO nanofiber mat, taken at different sample positions.

While in most applications, the nanofibers used have a very high surface-to-volume ratio, here the increase of the nanofiber diameter is not critical. Earlier investigations of the growth of Chinese hamster ovary (CHO) cells on different nanofiber mats revealed that growing on pure PAN nanofiber mats with their typical thin fiber diameters was impossible, while the cells grew well on PAN/gelatin nanofiber mats [36] with average fiber diameters of 490 ± 130 nm [37].

On the other hand, some of the fibers already show agglomerations of ZnO on the surface (Figure 1b). This indicates that the ZnO:PAN ratio chosen here, which is uncritical for electrospinning, may be too high and could be reduced in further experiments, depending on the desired application. It should be mentioned that this finding is similar to earlier investigations of PAN/TiO$_2$ nanofiber mats, which did not show increased fiber diameters, but also partly exhibited TiO$_2$ agglomerations along the fibers for a broad range of TiO$_2$ contents between 2.2% and 10.2% [38]. Former tests to reduce the TiO$_2$ agglomerations by longer stirring time, additional ultrasonic dispersion, reduced time between stirring and electrospinning, etc. indicated a general possibility to obtain a more homogeneous distribution of the TiO$_2$ and thus to reduce these agglomerations. Depending on the desired application—whether applications fixed along the fibers are undesired or even advantageous—similar experiments in combination with a variation of the ZnO content are necessary.

Next, it was tested whether the ZnO is freely available and could thus support the antibacterial activity of such samples [34]. Former investigations of PAN/TiO$_2$ nanofiber mats revealed very slow methylene blue degradation, indicating that only a small amount of TiO$_2$ is in contact with the environment [39]. Here, an easier method was used to investigate whether a significant amount of ZnO is in contact with the environment. ZnO belongs to the semiconductors used in dye-sensitized solar cells, typically showing a clearly visible bathochromic shift of the color of diverse dyes adsorbed on the material [40–43].

Figure 2 shows microscopic images of different raw and dyed fabrics. Figure 2a depicts a dyed cotton fabric, showing the typical color of the tea solution. The original PAN/ZnO nanofiber mat

(Figure 2b) shows white fibers. After dyeing such a nanofiber mat, drying and rinsing it with water, the nanofiber mat again looks white (Figure 2c). Obviously, not even the reddish color of the cotton fabric is reached. This is in strong contrast to dyeing a PAN/ZnO nanofiber mat that was previously dip-coated into an aqueous ZnO solution, showing the typical lilac color after the bathochromic shift due to the binding of the anthocyanins to the ZnO (Figure 2d), as it is also well-known from TiO$_2$ dyed with anthocyanins [44]. Apparently, no visible amount of ZnO could be dyed.

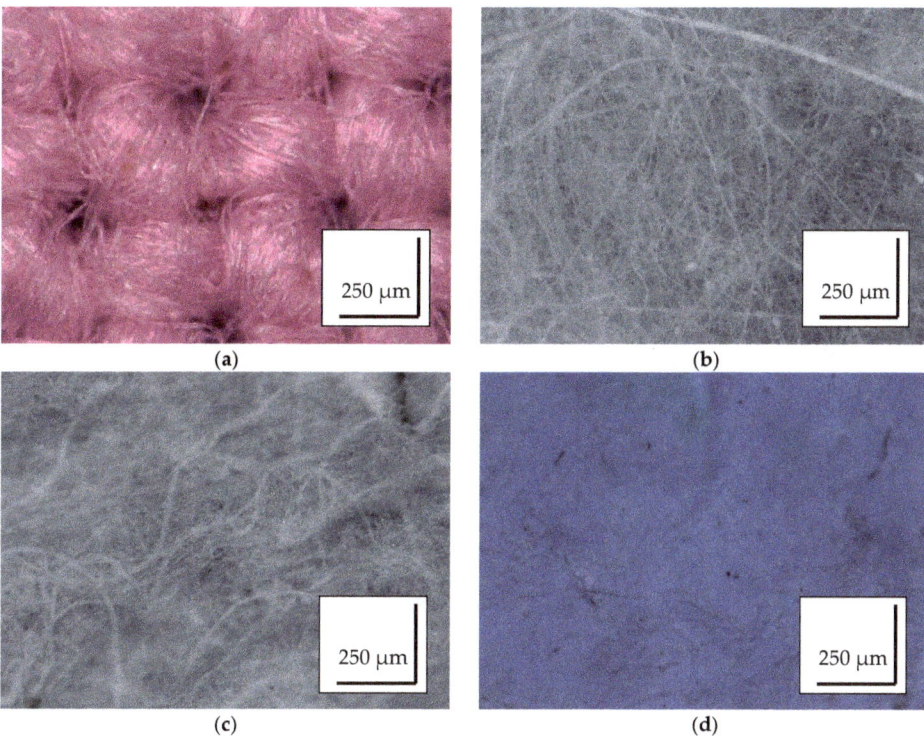

Figure 2. Microscopic images of (**a**) a cotton fabric dyed with anthocyanins; (**b**) a PAN/ZnO nanofiber mat; (**c**) a PAN/ZnO nanofiber mat dyed with anthocyanins; (**d**) a PAN/ZnO nanofiber mat dip-coated in ZnO and dyed with anthocyanins.

Depending on the desired application, it may be supportive to have a relatively inert fabric without the antibacterial activity of ZnO. However, for many applications, the antibacterial—or photocatalytic or other—properties of ZnO may be desired. Thus, it was tested whether ZnO nanoparticles could be coated on the PAN/ZnO fabric afterwards. While a previous experiment revealed the strong adhesion of TiO$_2$ nanoparticles in PAN nanofiber mats [44], the modified morphology of the PAN/ZnO nanofiber mat may also change the adhesion of nanoparticles coated on the fibers.

Unexpectedly, PAN/ZnO nanofibers enabled the even stronger adhesion of additional ZnO nanoparticles than pure PAN nanofiber mats. Figure 3 depicts both nanofiber mats, dip-coated in an aqueous ZnO solution, dried at ambient temperature and afterwards rinsed with water. It is clearly visible that more ZnO nanoparticles adhered to the PAN/ZnO nanofiber mat than to the PAN substrate. This is in contrast to coating PAN nanofiber mats with TiO$_2$, resulting in nano-composites that could even be applied as photo-electrodes in dye-sensitized solar cells [44].

Figure 3. SEM images of ZnO nanoparticles, applied by dip-coating on (**a**) a PAN nanofiber mat; (**b**) a PAN/ZnO nanofiber mat, after rinsing the substrates with water.

While the previous investigations concentrated on raw PAN/ZnO nanofiber mats, the main aim was the stabilization and carbonization of the electrospun samples. Figure 4 thus shows microscopic images of stabilized and carbonized PAN and PAN/ZnO nanofiber mats, respectively.

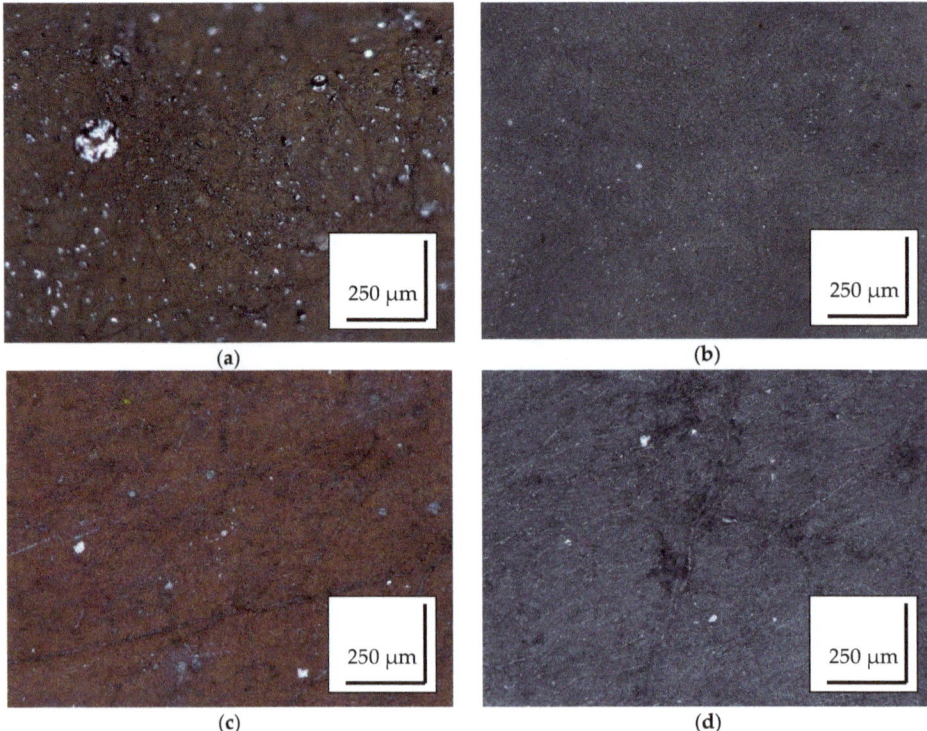

Figure 4. Microscopic images of (**a**) a stabilized PAN nanofiber mat; (**b**) a carbonized PAN nanofiber mat; (**c**) a stabilized PAN/ZnO nanofiber mat; (**d**) a carbonized PAN/ZnO nanofiber mat.

While the PAN fibers are too fine to be visible in these microscopic images, the stabilized and carbonized PAN/ZnO samples clearly show a fibrous structure. Carbonization at identical temperatures results in the more or less identical dark-grey color of the carbonized samples, while the brown colors of the stabilized samples differ slightly. The reddish brown of the stabilized PAN/ZnO nanofiber mat is similar to the color of the stabilized PAN/gelatin nanofibers, which has a similar diameter [37], suggesting that this color change may be attributed to diffraction at the nanofibers.

More importantly, the carbonized PAN/ZnO nanofiber mats are highly bendable. Figure 5 depicts carbonized nanofiber mats from pure electrospun PAN and PAN/ZnO nanofiber mats as precursors, respectively. The carbonized PAN nanofiber mat, stabilized and carbonized without fixation, breaks directly when grasped with the tweezers. Oppositely, the carbonized PAN/ZnO nanofiber mat, stabilized and carbonized under identical conditions, is bendable without any tendency to break, making this material more easily usable in applications such as 3D substrates for tissue engineering, where they are not used in a plane shape.

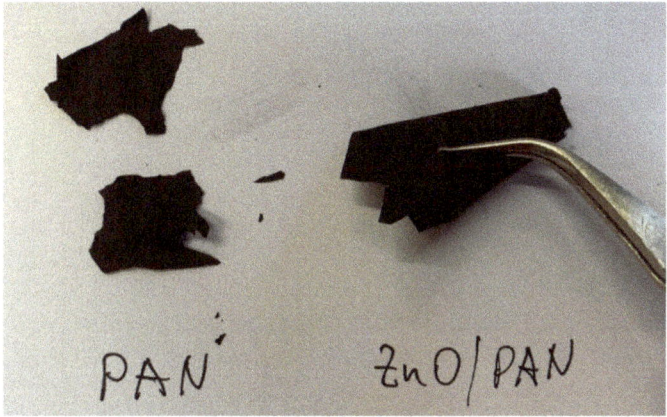

Figure 5. Bending a carbonized PAN nanofiber mat (stabilized without fixation) only slightly results in brittle failure (**left** side), while bending carbonized PAN/ZnO nanofiber mats (stabilized without fixation) is possible with a narrow bending radius (**right** side).

Finally, some other properties of the PAN/ZnO nanofiber mats are given. From the physical point of view, it is interesting that the carbon nanofiber mats produced from this precursor are not conductive (i.e., the sheet resistance is higher than 20 MΩ), opposite to PAN nanofiber mats carbonized under identical conditions [44].

For diverse applications, the carbon yield is an important factor, i.e., the ratio of the mass retained after carbonization to the mass of the original nanofiber mat. Typical values of the stabilization yield for a PAN nanofiber mat after stabilization with the process parameters applied here are in the range of 72%, the corresponding carbonization yield is approx. 29%, resulting in an overall mass yield of 21% [30].

For the PAN/ZnO nanofiber mats under investigation here, the stabilization yield is approx. 74%, while the carbonization yield is approx. 76%, resulting in an overall mass yield of 56%. While this seems to be a very promising result at first glance, it must be taken into account, nevertheless, that $\frac{1}{4}$ of the dry mass of the raw nanofiber mat consists of ZnO, which is thermally stable up to nearly 2000 °C and stays thus unaltered during carbonization at 500 °C. This reduces the overall carbon gain to (100% − 25%)/(56% − 25%) = 41%. This value is still approximately twice as high as the overall mass yield from a pure PAN nanofiber mat, indicating that future research on this topic is necessary to investigate the specific chemical or physical properties of the carbon/ZnO nanofiber mats that are necessary for different applications. Since this finding is unexpected, FTIR investigations were used to evaluate the chemical properties of the carbon/ZnO nanofiber mats.

Results for PAN/ZnO nanofiber mats in a raw state, after stabilization and carbonization, are depicted in Figure 6a. A ZnO absorption peak could be expected near 430 cm^{-1} [45], which cannot be achieved in the FTIR instrument used for this investigation. Annealing can be expected to reduce possible impurities in the ZnO so that no additional peaks should be visible in the spectra of the stabilized and the carbonized nanofiber mats [46].

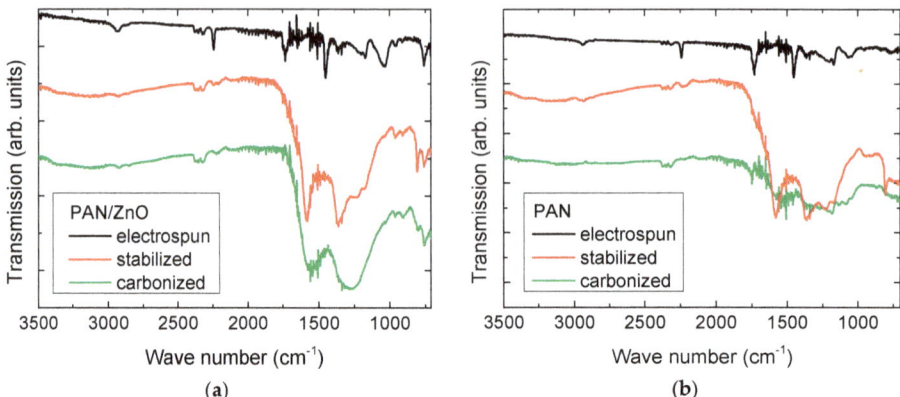

Figure 6. FTIR graphs of electrospun, stabilized and carbonized nanofiber mats from (**a**) PAN/ZnO; (**b**) PAN.

The spectra for raw and stabilized PAN/ZnO nanofiber mats indeed show the peaks expected for raw and stabilized PAN, as discussed in detail in [11,20]. The spectrum of the carbonized sample, however, shows clear residues of the stabilized material, indicating that carbonization is not completed here. For a better comparison, Figure 6b shows the corresponding spectra of a PAN nanofiber mat after electrospinning, stabilization and carbonization. While the first two spectra are qualitatively identical with those of PAN/ZnO, the latter shows clearly reduced features, underlining nearly complete carbonization.

This effect may be based on the increased nanofiber diameter, impeding sufficient heat supply to the fiber core. In other thicker nanofibers, like PAN/gelatin or PAN/poly(vinylidene fluoride), this problem may not occur since the blend partners are typically dissolved at temperatures below 500 °C [36,47], while PAN/TiO$_2$ nanofibers do not show an increased fiber diameter and thus behave similar to pure PAN fibers during carbonization [38]. The same effect of incomplete carbonization under identical conditions was already found for the carbonization of pure PAN nanofibers electrospun with a low voltage of 50 kV, resulting in thicker fibers, where no chemical difference to the usual PAN nanofibers electrospun with 80 kV, as done here, can be expected (to be published). This underlines that optimal carbonization parameters do not only have to be distinguished between nanofibers and microfibers, but also between differently electrospun nanofibers with correspondingly different diameters.

The FTIR graphs clearly show that a deeper investigation of the carbonization process and possibly even an optimization of the stabilization step are necessary to create carbon/ZnO nanofiber mats with good mechanical and reliable chemical properties.

4. Conclusions

PAN/ZnO nanofiber mats were prepared by electrospinning, resulting in nanofibers with diameters approximately three times the value of pure PAN nanofibers. While the ZnO included in the polymer matrix was not in contact with the environment, as shown by a simple dyeing test, additional ZnO nanoparticles adhered better on these nanofiber mats than on those from pure PAN, enabling the addition of antibacterial or other desired properties of ZnO.

Stabilization and carbonization with the process parameters typical for PAN nanofiber mats resulted in highly flexible nanofiber mats. FTIR investigations revealed that carbonization was not completed for the PAN/ZnO nanofiber mats, suggesting further investigation and optimization of the carbonization, and possibly also of the stabilization, process is needed to prepare flexible carbon/ZnO nanofiber mats with well-defined chemical properties as 3D substrates in tissue engineering.

Author Contributions: Conceptualization, T.B., L.S. and A.E.; formal analysis, L.S., A.E. and T.B.; investigation, M.T., A.M., M.K., L.S. and C.G.; writing—original draft preparation, A.E.; writing—review and editing, all authors; visualization, A.E.; supervision, A.E. and T.B.

Funding: This project was funded by the Erasmus+ program of the European Union, by the HiF funds of Bielefeld University of Applied Sciences, by the Deutsche Bundesstiftung Umwelt DBU (German Federal Environmental Foundation) and by the Silesian University of Technology (SUT) Rector Grant 14/990/RGJ18/0099. This article was funded by the Open Access Publication Fund of Bielefeld University of Applied Sciences and the Deutsche Forschungsgemeinschaft (DFG, German Research Foundation)—414001623.

Conflicts of Interest: The authors declare no conflict of interest. The funders had no role in the design of the study; in the collection, analyses, or interpretation of data; in the writing of the manuscript; or in the decision to publish the results.

References

1. Greiner, A.; Wendorff, J.H. Electrospinning: A fascinating method for the preparation of ultrathin fibers. *Angew. Chem. Int. Ed.* **2007**, *46*, 5670–5703. [CrossRef] [PubMed]
2. Lv, D.; Wang, R.X.; Tang, G.S.; Mou, Z.P.; Lei, J.D.; Ha, J.Q.; de Smedt, S.; Xiong, R.H. Correction to Ecofriendly Electrospun Membranes Loaded with Visible-Light-Responding Nanoparticles for Multifunctional Usages: Highly Efficient Air Filtration, Dye Scavenging, and Bactericidal Activity. *ACS Appl. Mater. Interfaces* **2019**, *11*, 12880–12889. [CrossRef] [PubMed]
3. Zhu, M.M.; Han, J.Q.; Wang, F.; Shao, W.; Xiong, R.H.; Zhang, Q.L.; Pan, H.; Yang, Y.; Samal, S.K.; Zhang, F.; et al. Electrospun Nanofibers Membranes for Effective Air Filtration. *Macromol. Mater. Eng.* **2017**, *302*. [CrossRef]
4. Lv, D.; Zhu, M.M.; Jiang, Z.C.; Jiang, S.H.; Zhang, Q.L.; Xiong, R.H.; Huang, C.B. Green Electrospun Nanofibers and Their Application in Air Filtration. *Macromol. Mater. Eng.* **2018**, *303*. [CrossRef]
5. Qin, X.-H.; Wang, S.-Y. Electrospun nanofibers from crosslinked poly(vinyl alcohol) and its filtration efficiency. *Appl. Polym. Sci.* **2008**, *109*, 951–956. [CrossRef]
6. Roche, R.; Yalcinkaya, F. Incorporation of PVDF nanofibre multilayers into functional structure for filtration applications. *Nanomaterials* **2018**, *8*, 771. [CrossRef]
7. Roche, R.; Yalcinkaya, F. Electrospun polyacrylonitrile nanofibrous membranes for point-of-use water and air cleaning. *ChemistryOpen* **2019**, *8*, 97–103. [CrossRef]
8. Fu, Q.S.; Lin, G.; Chen, X.D.; Yu, Z.X.; Yang, R.S.; Li, M.T.; Zeng, X.G.; Chen, J. Mechanically Reinforced PVdF/PMMA/SiO$_2$ Composite Membrane and Its Electrochemical Properties as a Separator in Lithium-Ion Batteries. *Energy Technol.* **2018**, *6*, 144–152. [CrossRef]
9. Gao, S.T.; Tang, G.S.; Hua, D.W.; Xiong, R.H.; Han, J.Q.; Jiang, S.H.; Zhang, Q.L.; Huang, C.B. Stimuli-responsive bio-based polymeric systems and their applications. *J. Mater. Chem. B* **2019**, *7*, 709–729. [CrossRef]
10. Sabantina, L.; Kinzel, F.; Hauser, T.; Többer, A.; Klöcker, M.; Döpke, C.; Böttjer, R.; Wehlage, D.; Rattenholl, A.; Ehrmann, A. Comparative study of Pleurotus ostreatus mushroom grown on modified PAN nanofiber mats. *Nanomaterials* **2019**, *9*, 475. [CrossRef]
11. Klinkhammer, K.; Seiler, N.; Grafahrend, D.; Gerardo-Nava, j.; Mey, J.; Brook, G.A.; Möller, M.; Dalton, P.D.; Klee, D. Deposition of electrospun fibers on reactive substrates for In Vitro investigations. *Tissue Eng. Part C* **2009**, *15*, 77–85. [CrossRef] [PubMed]
12. Rahaman, M.S.A.; Ismail, A.F.; Mustafa, A. A review of heat treatment on polyacrylonitrile fiber. *Polym. Degrad. Stab.* **2007**, *92*, 1421–1432. [CrossRef]
13. Sabantina, L.; Mirasol, J.R.; Cordero, T.; Finsterbusch, K.; Ehrmann, A. Investigation of Needleless Electrospun PAN Nanofiber Mats. *AIP Conf. Proc.* **2018**, *1952*. [CrossRef]

14. Grothe, T.; Wehlage, D.; Böhm, T.; Remche, A.; Ehrmann, A. Needleless electrospinning of PAN nanofibre mats. *Tekstilec* **2017**, *60*, 290–295. [CrossRef]
15. Chen, J.; Harrison, I. Modification of polyacrylonitrile (PAN) carbon fiber precursor via postspinning plasticization and stretching in dimethyl formamide (DMF). *Carbon* **2002**, *40*, 25–45. [CrossRef]
16. Bashir, Z. A critical review of the stabilisation of polyacrylonitrile. *Carbon* **1991**, *29*, 1081–1090. [CrossRef]
17. Dalton, S.; Heatley, F.; Budd, P.M. Thermal stabilization of polyacrylonitrile fibres. *Polymer* **1999**, *40*, 5531–5543. [CrossRef]
18. Ismar, E.; Sezai Sarac, A. Oxidation of polyacrylonitrile nanofiber webs as a precursor for carbon nanofiber: Aligned and non-aligned nanofibers. *Polym. Bull.* **2017**, *75*, 485–499. [CrossRef]
19. Fitzer, E.; Frohs, W.; Heine, M. Optimization of stabilization and carbonization treatment of PAN fibres and structural characterization of the resulting carbon fibres. *Carbon* **1986**, *24*, 387–395. [CrossRef]
20. Mólnar, K.; Szolnoki, B.; Toldy, A.; Vas, L.M. Thermochemical stabilization and analysis of continuously electrospun nanofibers. *J. Therm. Anal. Calorim.* **2014**, *117*, 1123–1135. [CrossRef]
21. Sabantina, L.; Klöcker, M.; Wortmann, M.; Rodrígues-Mirasol, J.; Cordero, T.; Moritzer, E.; Finsterbusch, K.; Ehrmann, A. Stabilization of PAN nanofiber mats obtained by needleless electrospinning using DMSO as solvent. *J. Ind. Text.* **2018**. [CrossRef]
22. Mathur, R.; Bahl, O.; Mittal, J. A new approach to thermal stabilization of PAN fibres. *Carbon* **1992**, *30*, 657–663. [CrossRef]
23. Moon, S.C.; Farris, R.J. Strong electrospun nanometer-diameter polyacrylonitrile carbon fiber yarns. *Carbon* **2009**, *47*, 2829–2839. [CrossRef]
24. Wu, S.; Zhang, F.; Yu, Y.H.; Li, P.; Yang, X.P.; Lu, J.G.; Rye, S.K. Preparation of PAN-based carbon nanofibers by hot-stretching. *Compos. Interfaces* **2008**, *15*, 671–677. [CrossRef]
25. Xie, Z.; Niu, H.; Lin, T. Continuous polyacrylonitrile nanofiber yarns: Preparation and drydrawing treatment for carbon nanofiber production. *RSC Adv.* **2015**, *5*, 15147–15153. [CrossRef]
26. Ma, S.; Liu, J.; Liu, Q.; Liang, J.Y.; Zhao, Y.; Fong, H. Investigation of structural conversion and size effect from stretched bundle of electrospun polyacrylonitrile copolymer nanofibers during oxidative stabilization. *Mater. Des.* **2016**, *95*, 387–397. [CrossRef]
27. Ma, S.; Liu, J.; Qu, M.; Wang, X.; Huang, R.; Liang, J. Effects of carbonization tension on the structural and tensile properties of continuous bundles of highly aligned electrospun carbon nanofibers. *Mater. Lett.* **2016**, *183*, 369–373. [CrossRef]
28. Wu, M.; Wang, Q.Y.; Li, K.; Wu, Y.Q.; Liu, H.Q. Optimization of stabilization conditions for electrospun polyacrylonitrile nanofibers. *Polym. Degrad. Stab.* **2012**, *97*, 1511–1519. [CrossRef]
29. Santos de Oliveira, M.J.; Manzolli Rodrigues, B.V.; Marcuzzo, J.S.; Guerrini, L.M.; Baldan, M.R.; Rezende, M.C. A statistical approach to evaluate the oxidative process of electrospun polyacrylonitrile ultrathin fibers. *J. Appl. Polym. Sci.* **2017**, *134*. [CrossRef]
30. Sabantina, L.; Wehlage, D.; Klöcker, M.; Mamun, A.; Grothe, T.; Rodrígues Mirasol, J.; Cordero, T.; Finsterbusch, K.; Ehrmann, A. Stabilization of electrospun PAN/gelatin nanofiber mats for carbonization. *J. Nanomater.* **2018**, *2018*. [CrossRef]
31. Sabantina, L.; Ángel Rodríguez-Cano, M.; Klöcker, M.; García-Mateos, F.J.; Ternero-Hidalgo, J.J.; Mamun, A.; Beermann, F.; Schwakenberg, M.; Voigt, A.-L.; Rodríguez-Mirasol, J.; et al. Fixing PAN nanofiber mats during stabilization for carbonization and creating novel metal/carbon composites. *Polymers* **2018**, *10*, 735. [CrossRef] [PubMed]
32. Kim, J.H.; Mirzaei, A.; Kim, H.W.; Wu, P.; Kim, S.S. Design of supersensitive and selective ZnO-nanofiber-based sensors for H2 gas sensing by electron-beam irradiation. *Sens. Actuators B Chem.* **2019**, *293*, 210–223. [CrossRef]
33. Yun, S.I.; Kim, S.H.; Kim, D.W.; Kim, Y.A.; Kim, B.H. Facile preperation and capacitive properties of low-cost carbon nanofibers with ZnO derived lignin and pitch as supercapacitor electrodes. *Carbon* **2019**, *149*, 637–645. [CrossRef]
34. Chen, Y.; Lu, W.P.; Guo, Y.C.; Zhu, Y.; Song, Y.P. Electrospun gelatin fibers surface loaded ZnO particles as a potential biodegradable antibacterial wound dressing. *Nanomaterials* **2019**, *9*, 525. [CrossRef] [PubMed]
35. Mamun, A. Review of possible applications of nanofibrous mats for wound dressings. *Tekstilec* **2019**, *62*, 89–100. [CrossRef]

36. Wehlage, D.; Blattner, H.; Sabantina, L.; Böttjer, R.; Grothe, T.; Rattenholl, A.; Gudermann, F.; Lütkemeyer, D.; Ehrmann, A. Sterilization of PAN/gelatine nanofibrous mats for cell growth. *Tekstilec* **2019**, *62*, 78–88. [CrossRef]
37. Wehlage, D.; Böttjer, R.; Grothe, T.; Ehrmann, A. Electrospinning water-soluble/insoluble polymer blends. *AIMS Mater. Sci.* **2018**, *5*, 190–200. [CrossRef]
38. Sabantina, L.; Böttjer, R.; Wehlage, D.; Grothe, T.; Klöcker, M.; García-Mateos, F.J.; Rodríguez-Mirasol, J.; Cordero, T.; Ehrmann, A. Morphological study of stabilization and carbonization of polyacrylonitrile/TiO$_2$ nanofiber mats. *J. Eng. Fibers Fabr.* **2019**, *14*. [CrossRef]
39. Grothe, T.; Böttjer, R.; Wehlage, D.; Großerhode, C.; Storck, J.L.; Juhász Junger, I.; Mahltig, B.; Grethe, T.; Graßmann, C.; Schwarz-Pfeiffer, A.; et al. Photocatalytic properties of TiO$_2$ composite nanofibers electrospun with different polymers. In *Inorganic and Composite Fibers*; Mahltig, B., Kyosev, Y., Eds.; Woodhead Publishing: Cambridge, UK, 2018; pp. 303–319.
40. Dentani, T.; Kubota, Y.; Funabiki, K.; Jin, J.; Yoshida, T.; Minoura, H.; Miura, H.; Matsui, M. Novel thiopene-confugated indoline dyes for zinc oxide solar cells. *New J. Chem.* **2009**, *33*, 93–101. [CrossRef]
41. Salih, E.Y.; Sabri, M.F.M.; Tan, S.T.; Sulaiman, K.; Hussein, M.Z.; Said, S.M.; Yap, C.C. Preparation and characterization of ZnO/ZnAl$_2$O$_4$-mixed metal oxides for dye-sensitized photodetector using Zn/Al-layered double hydroxide as precursor. *J. Nanoparticle Res.* **2019**, *21*, 55. [CrossRef]
42. Saboor, A.; Shah, S.M.; Hussain, H. Band gap tuning and applications of ZnO nanorods in hybrid solar cell: Ag-doped verses Nd-doped ZnO nanorods. *Mater. Sci. Semicond. Process.* **2019**, *93*, 215–225. [CrossRef]
43. Sehgal, P.; Narula, A.K. Metal substituted metalloporphyrins as efficient photosensitizers for enhanced solar energy conversion. *J. Photochem. Photobiol. A Chem.* **2019**, *375*, 91–99. [CrossRef]
44. Mamun, A.; Trabelsi, M.; Klöcker, M.; Sabantina, L.; Großerhode, C.; Blachowicz, T.; Grötsch, G.; Cornelißen, C.; Streitenberg, A.; Ehrmann, A. Electrospun nanofiber mats with embedded non-sintered TiO$_2$ for dye sensitized solar cells (DSSCs). *Fibers* **2019**, *7*, 60. [CrossRef]
45. Becheri, A.; Dürr, M.; Lo Nostro, P.; Baglioni, P. Synthesis and characterization of zinc oxide nanoparticles: Application to textiles as UV-absorbers. *J. Nanoparticle Res.* **2008**, *10*, 679–689. [CrossRef]
46. Xiong, G.; Pal, U.; Serrano, J.G.; Ucer, K.B.; Williams, R.T. Photoluminescence and FTIR study of ZnO nanoparticles: The impurity and defect perspective. *Phys. Status Solidi* **2006**, *10*, 3577–3581. [CrossRef]
47. Mamun, A.; Trabelsi, M.; Wortmann, M.; Frese, N.; Moritzer, E.; Gölzhäuser, A.; Hüsgen, B.; Sabantina, L. Stabilization and carbonization of PAN/PVDF nanofiber blend scaffolds. In preparation.

© 2019 by the authors. Licensee MDPI, Basel, Switzerland. This article is an open access article distributed under the terms and conditions of the Creative Commons Attribution (CC BY) license (http://creativecommons.org/licenses/by/4.0/).

Article

Shape-Memory Nanofiber Meshes with Programmable Cell Orientation

Eri Niiyama [1,2], Kanta Tanabe [1,3], Koichiro Uto [4], Akihiko Kikuchi [5] and Mitsuhiro Ebara [1,2,3,*]

1. International Center for Materials Nanoarchitectonics (WPI-MANA), National Institute for Materials Science (NIMS), Tsukuba, Ibaraki 305-0044, Japan; s1630131@u.tsukuba.ac.jp (E.N.); tanabe19331510@yahoo.co.jp (K.T.)
2. Graduate School of Pure and Applied Sciences, University of Tsukuba, Tsukuba, Ibaraki 305-8577, Japan
3. Graduate School of Industrial Science and Technology, Tokyo University of Science, Katsushika, Tokyo 125-8585, Japan
4. International Center for Young Scientists (ICYS), National Institute for Materials Science (NIMS), Tsukuba, Ibaraki 305-0044, Japan; uto.koichiro@nims.go.jp
5. Department of Materials Science and Technology, Tokyo University of Science, Katsushika, Tokyo 125-8585, Japan; kikuchia@rs.noda.tus.ac.jp
* Correspondence: ebara.mitsuhiro@nims.go.jp

Received: 14 February 2019; Accepted: 25 February 2019; Published: 1 March 2019

Abstract: In this work we report the rational design of temperature-responsive nanofiber meshes with shape-memory properties. Meshes were fabricated by electrospinning poly(ε-caprolactone) (PCL)-based polyurethane with varying ratios of soft (PCL diol) and hard [hexamethylene diisocyanate (HDI)/1,4-butanediol (BD)] segments. By altering the PCL diol:HDI:BD molar ratio both shape-memory properties and mechanical properties could be readily turned and modulated. Though mechanical properties improved by increasing the hard to soft segment ratio, optimal shape-memory properties were obtained using a PCL/HDI/BD molar ratio of 1:4:3. Microscopically, the original nanofibrous structure could be deformed into and maintained in a temporary shape and later recover its original structure upon reheating. Even when deformed by 400%, a recovery rate of >89% was observed. Implementation of these shape memory nanofiber meshes as cell culture platforms revealed the unique ability to alter human mesenchymal stem cell alignment and orientation. Due to their biocompatible nature, temperature-responsivity, and ability to control cell alignment, we believe that these meshes may demonstrate great promise as biomedical applications.

Keywords: shape memory nanofiber; shape memory polymer; poly(ε-caprolactone); melting temperature; cell orientation; polyurethane

1. Introduction

Stimuli-responsive fibrous materials with shape-memory properties (also called "shape-memory fibers") have received great attention for their potential regenerative medicine, filtration, robotics, and catalysis applications [1–8]. Shape-memory polymers (SMPs) are a class of temperature-responsive materials that can change from a temporary shape to a memorized permanent shape upon the application of heat. Both glass transition temperature (T_g) and melting temperature (T_m) have been leveraged to initiate shape-switching; however, T_m is often favored in the design of SMPs because the enthalpy changes of the solid–liquid phase transition are much larger than that of the glass–rubber transition or liquid crystalline transition. While many SMPs have been explored, poly(ε-caprolactone) (PCL)-based SMPs have been utilized as a biomaterial due to their well-characterized biocompatibility [9] and biodegradability [10,11]. Combining SMPs with various fabrication processes has resulted in diverse structures including shape-memory films, foams, particles, surfaces, and fibers [4,12–19]. Particularly, SMPs in the form of fibers are generating great interest in structural and functional applications owing to their extremely high surface area, porous structure, and filtration/penetration properties [20–24]. Various thermoplastic polymers have been spun into nanofibers by the electrospinning method, which has been extensively acknowledged as an efficient and convenient approach for producing nanofibrous materials [24,25]. Fibrous materials are attractive as tissue engineering scaffolds because of their ability to enhance cell attachment, control pore architecture and create a 3-D microenvironment that encourages cell–cell contact [26,27]. Fibrous scaffolds have been implemented in cardiovascular [28–30], musculoskeletal [31,32], neural [33], and stem cell tissue engineering [34,35]. In addition, submicron-diameter fibers can provide tissue-matching mechanical compliance and provide topographic cues similar to that of native extracellular matrices (ECM). Studies have revealed that ECM topography greatly alters cell differentiation and tissue function [36–39].

Numerous studies have demonstrated the processable and structural advantages of nanofibrous materials with shape-memory properties. Matsumoto et al. using poly(ω-pentadecalactone) and PCL shape-memory microfibers achieved a strain recovery rate (R_r) of >89% and a strain fixity rate (R_f) of >82% by applying small deformations (~25%) [40]. Fejős et al. generated triple-shape memory nanofibers to study the effect of the structure on triple-shape memory by exploiting the T_g of epoxy and the T_m of PCL as the shape-switching temperatures [41]. Good strain recovery rates (>94% for T_g and >89% for T_m) were obtained; however the samples were deformed to only 2% strain. Ji et al. developed a series of shape-memory polyurethanes with varying soft-hard segment ratios [42]. The R_f increased from 75% to 92% and the R_r decreased from 92% to 85% with increasing hard segment contents. Barmouz et al. investigated the shape memory behavior of poly(lactic acid)-thermoplastic polyurethane/cellulose-nanofiber bio-nanocomposites. They concluded that through the addition of cellulose nanofibers, stress recovery of >40% could be achieved with little change in strain recovery [43]. Kawaguchi et al. found that by combining chitosan fibers with polyether-based thermoplastic polyurethane, the crystal structure gradually changed from semi-crystalline to amorphous state despite little or no change to the glass transition temperature. The elastic module of this hybrid material increased by 40% as compared with pure thermoplastic polyurethane. Shape recovery of these materials could be achieved at temperatures ranging from 25 to 70 °C. [44]. Aslan et al. reported that shape polyurethane fibers prepared by wet spinning demonstrated R_f and R_r of 71% and 91%, respectively [45]. Although these values are considered good for the shape-memory behavior, shape-memory nanofibers generally show relatively lower shape fixity and recovery properties as compared with shape-memory films. One of the major reasons is its polymer network architecture. In general, SMP systems can be broadly classified into two types based on the network architecture: (1) a physically cross-linked network and (2) a covalently cross-linked network [46]. Because it is difficult to cross-link the networks chemically during electrospinning, physical cross-linking is more suitable for electrospun nanofiber systems. However, compared with chemical cross-linking, physical cross-linking generally results in less structurally stable networks.

In this study, we describe the rational design of shape-memory nanofiber meshes generated by electrospinning PCL-based polyurethane which demonstrate higher shape-memory abilities. A series of polyurethanes with different ratios of soft and hard segments were prepared. Hard segments participate in hydrogen bonding and crystallization conferring rigidity, while, soft segments demonstrate a reversible phase transformation at the T_m, conferring shape-memory properties. Altering the ratio of hard and soft segments in a single mesh resulted in dramatic differences in fiber processability, mechanical properties, T_m, and fiber stability. Systematic variations of these parameters resulted in deformations greater than or equal to 400%. In addition, we examined the control of cell orientation on the nanofiber meshes with different fiber alignments because controlling cell alignment is one of the most crucial steps to creating practical tissue scaffolds such as cardiovascular, musculoskeletal, neural areas because many cells in these tissues align well along the ECM. Also, recent studies in mechanobiology field have revealed that ECM topography greatly alter cell differentiation and tissue function [14]. The polymer design strategy applied in this study can increase the prospective applications of shape-memory fibers in the biomedical field.

2. Materials and Methods

2.1. Materials

HDI, BD, ε-caprolactone (CL), and 1,1,1,3,3,3-hexafluoro-2-propanol (HFIP) were obtained from Tokyo Chemical Industry Co., Ltd. (Tokyo, Japan). Tin (II) 2-ethylhexanoate, rhodamine phalloidin, 4',6-diamidine-2-phenylindole dihydrochloride (DAPI), fibronectin, and 0.1% triton X-100 were purchased from Sigma-Aldrich Japan (Tokyo, Japan). Phosphate buffered saline (PBS) was purchased from Nakalai Tesque (Kyoto, Japan). Xylene, *n*-hexane, diethylether, tetrahydrofuran (THF), and paraformaldehyde were obtained from FUJIFILM Wako Pure Chemical Corporation (Osaka, Japan). Human mesenchymal stem cells (hMSCs) were purchased from Lonza (Basel, Switzerland).

2.2. Polymerization and Characterization

BD (194 µL, 2.2 mmol) as an initiator was dried in vacuum overnight in a 300 mL round-bottom flask. CL (46.3 mL, 0.44 mol) was added into the flask and stirred under a N_2 atmosphere. Five drops of tin (II) 2-ethylhexanoate as a catalyst (5 droplets, 0.5 mmol) was then dropwise added and stirred at 120 °C for 24 h. The product was then completely dissolved in THF. The obtained PCL was purified by reprecipitation from hexane and diethyl ether. Then, 800 mg of purified PCL and 20 mL of xylene were added into a 50 mL sample tube and stirred at 60 °C for 15 min. HDI and tin (II) 2-ethylhexanoate were then added into the mixture and stirred at 60 °C for 30 min followed by the addition of BD. The added amounts of HDI and BD were 44 µL (0.27 mmol) and 16 µL (0.18 mmol), 58 µL (0.36 mmol) and 24 µL (0.27 mmol), and 72.5 µL (0.45 mmol) and 32 µL (0.36 mmol) to obtain PCL:HDI:BD molar ratios of 1:3:2, 1:4:3, and 1:5:4 (soft:hard segment = 1:5, 1:7, and 1:9), respectively. After stirring at 60 °C for 3 h, the mixture was purified by reprecipitation with a mixed solution of hexane and chloroform (80:3). The obtained polymer was filtered and dried in vacuum. The structures were determined by ^{1}H-nuclear magnetic resonance (NMR) spectroscopy (JEOL, Tokyo, Japan) with CDCl$_3$ as a solvent (Figure S1). Urethane bonds were analyzed by Fourier transform-infrared spectroscopy (FT-IR) (JEOL, Tokyo, Japan) with KBr pellet (Figure S2). The molecular weights were determined by Gel permeation chromatography (GPC) equipped with TSKgel G4000Hhr and TSKgel G3000Hhr columns and a refractive detector using N,N-dimethylformamide (DMF) with 10 mM LiCl as the eluent and solvent (0.8 mL/min, 40 °C) (HLC-8220GPC, Tosho Corporation, Tokyo, Japan) (Table 1).

Table 1. Characteristic data of a series of PCL-based polyurethanes. PDI: Polydispersity index.

Samples	Composition (Molar Ratio)	Segment Ratio [1] (Molar Ratio)		Segment Ratio (w/w%)		Feed			Molecular Weight [2]		PDI [2]
	PCL:HDI:BD	Soft	Hard	Soft	Hard	PCL (mg)	HDI (µL)	BD (µL)	M_w	M_n	(M_w/M_n)
PCL	-	-	-	-	-	-	-	-	59,700	46,300	1.29
PCL-6.8	1:3:2	1	5	93.2	6.8	800	44	16	78,900	54,000	1.46
PCL-9.2	1:4:3	1	7	90.8	9.2	800	58	24	73,200	53,000	1.38
PCL-11.3	1:5:4	1	9	88.7	11.3	800	73	32	93,500	68,000	1.37

[1] Soft:Hard = PCL:HDI + BD (PCL; poly(ε-caprolactone), HDI; hexamethylene diisocyanate, BD;1,4-butanediol).
[2] Measured by GPC.

2.3. Electrospinning Method and Characterization of Nanofibers

The electrospinning solution was prepared by dissolving the polymer in HFIP (40 w/v%). PCL with different ratios of HDI and BD were electrospun into nanofibers using an applied voltage of 25 kV, a needle gauge of 23, a flow rate of 0.5 mL/h, and a 15 cm separation between the needle and the collector plate (Nanon-01A, MECC Co., Ltd., Fukuoka, Japan) (n = 3). The formation of electrospun nanofibers was observed using a scanning electron microscope (SEM; SU8000, Hitachi High-Technologies Corporation, Tokyo, Japan). The thermal property of the nanofibers with different ratios of HDI and BD was measured by differential scanning calorimetry (DSC; 6100, SEIKO Instruments, Chiba, Japan) at a heating/cooling rate of 5 °C/min. For the thermal stability test, a nanofiber mesh with a soft:hard segment ratio of 1:5, 1:7, or 1:9 was placed in an oven at 60 °C for 24 h. The nanofiber morphology before and after heating was compared by SEM observation. Because polyurethanes are known as hygroscopic material, we have conducted all experiments under constant temperature/humidity conditions (21 °C/25%).

2.4. Shape Memory Behavior

The shape memory effect of the electrospun nanofibers with a soft:hard segment ratio of 1:7 was evaluated in terms of morphology, diameter change, orientation, and shape recovery rate of the nanofibers before deformation, after deformation, and after shape memory recovery. The nanofiber was first heated at 60 °C in water and then stretched to a temporary shape. The nanofiber was reheated at 60 °C, which led to shape recovery. The formation of nanofibers and their morphologies before deformation, during formation of the temporary shape, and after shape recovery were observed by SEM. The surface and cross-section of the shape memory fiber were observed. From the magnified cross-sectional and top-view images, the diameters of the nanofiber before deformation, during formation of the temporary shape, and after shape recovery were calculated. The orientation of the nanofibers before and after deformation, and after shape recovery was analyzed, and the orientation images were created by Image J software (Image J, the National Institutes of Health, Bethesda, MD, USA) using the plugin orientation J to obtain direction distribution maps and their histograms (Figure S3). The R_r and R_f were calculated by cutting the shape memory nanofiber to 1 cm length and comparing the length of the nanofiber before deformation and shape recovery (n = 3). The strain fixity rate describes the ability to fix the mechanical deformation as $R_f(N) = \varepsilon_f(N) / \varepsilon_m \times 100\%$. The strain recovery rate was the ability of the material to recover its permanent shape, which calculated from $R_r = (\varepsilon_m - \varepsilon_r(N)) / \varepsilon_m - \varepsilon_r(N-1) \times 100\%$ (ε_m; max strain, ε_r; recovered strain, ε_f; final strain after deformation, N is the number of cycles). The nanofiber was heated at 60 °C in water, and then stretched to a certain elongation degree (200%, 300%, or 400%). The nanofiber was again placed in water at 60 °C for shape memory recovery. The recovery stress of PCL-9.2 nanofiber was also measured through a tensile test (EZ-S, SHIMADZU, Kyoto, Japan). At first, the sample was extended to 150% elongation over 60 °C in chamber (M-600FN, TAITEC, Saitama, Japan) and then, the sample was cooled down to 0 °C. The sample was maintained at the constant deformation for 1 hour. Subsequently, the sample was reheated to 60 °C and the stress stored in the sample was released. The largest value was taken as the representative of the recovery stress at this strain (Figure S4).

2.5. Cell Culture on Nanofiber Mesh

Before starting the hMSC culture, the original and temporarily stretched electrospun nanofibers with a soft:hard segment ratio of 1:7 were sterilized by ultraviolet (UV) irradiation for 15 min. The nanofibers were coated with 20 µg mL^{-1} fibronectin for 1 h at 37 °C. After washing with PBS, hMSCs were seeded at a density of 5000 cells cm^{-2} on the sterilized nanofiber and cultured in a hMSC growth medium (MSCGMTM, Lonza, Basel, Switzerland)) for 1 day. The hMSCs cultured on the nanofibers were fixed using 4% paraformaldehyde for 15 min and then permeabilized using 0.1% Triton X-100 for 5 min. Cells were stained using rhodamine phalloidin and DAPI to visualize F-actin and nuclei respectively. Finally, the cell morphology on the nanofiber meshes was imaged using a fluorescence microscope (ECLIPSE Ti2, Nikon, Tokyo, Japan). The phase contrast and fluorescence images were taken for cells and nanofibers in the same field. Finally, three images including phase contrast (nanofibers and cells), fluorescence images for nucleus (DAPI) and F-actins (Phalloidin) from the same field were merged. To investigate the overview of cellular alignment, large field scan was also performed. All original images were shown in Figure S5. Cell viability on the nanofiber meshes before and after deformation was evaluated using Alamar blue. The metabolic activity of hMSCs cultured on nanofiber meshes before and after deformation as well as on glass substrates was assessed (n = 4). Briefly, cells were treated with Alamar blue reagent diluted in culture media (10%) for 4 hours at 37 °C. Media was then sampled and analyzed using a flourescence plate reader (PerkinElmer Co., Ltd., Kanagawa, Japan).

3. Results

3.1. Fabrication of Poly(ε-Caprolactone) PCL-Based Polyurethane Nanofiber Meshes

Three different PCL-based polyurethanes were synthesized by reacting a PCL diol (generated by a caprolactone ring opening polymerization [2,13,15]), 1,4-butanediol (BD) and hexamethylene diisocyanate (HDI) (Scheme 1) in varying soft:hard segment ratios (1:5, 1:7, and 1:9 corresponding to PCL:HDI:BD molar ratios of 1:3:2, 1:4:3, and 1:5:4, respectively). Polymer synthesis was then confirmed by ^{1}H-NMR spectroscopy and gel permeation chromatography (GPC) (Figure S1 and Table 1). Soft:hard segment ratios of 1:5, 1:7, and 1:9 resulted in polymers containing roughly 6.8, 9.2, and 11.3 w/w% of hard segments and were thus abbreviated PCL-6.8, PCL-9.2 and PCL-11.3. Urethane peaks were detected from ^{1}H-NMR spectrum at 3.2 ppm and FT-IR spectrum as C–N stretching and N–H bending at 1540 cm^{-1}, bending vibrations of the N–H bond at 1580 cm^{-1}, and free amide group (–NH–) peak at 3340 cm^{-1} (Figure S1 and S2) [47].

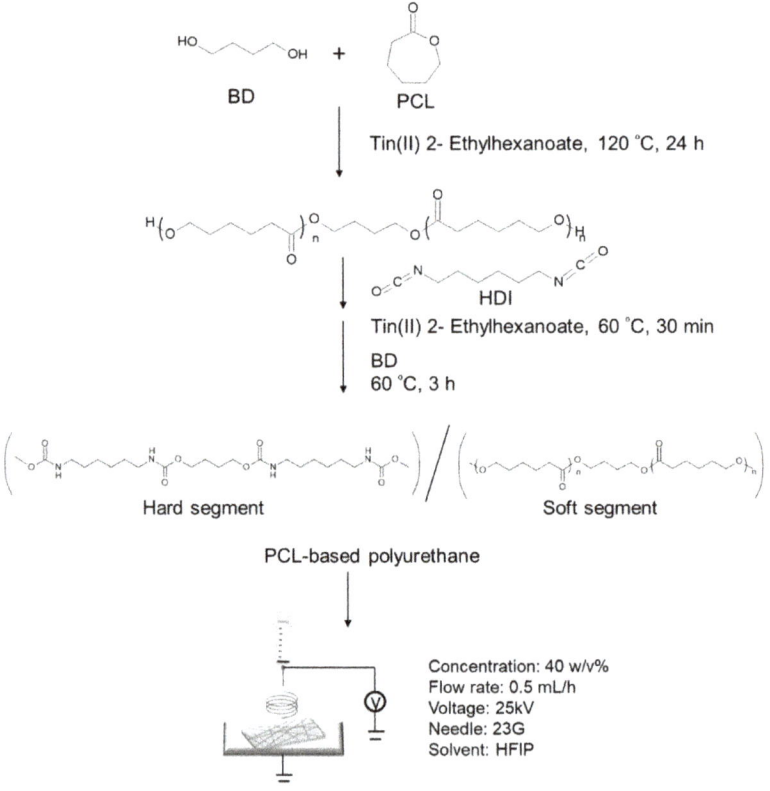

Scheme 1. Synthesis of poly(ε-caprolactone) (PCL)-based polyurethanes and fabrication of nanofiber mesh by electrospinning.

As molecular weight is a critical factor for electrospinning and intermolecular chain entanglement is necessary, number-average molecular weight (M_n) was closely monitored. Figure S6 shows the morphology of the electrospun PCL with a molecular weight of 5000. Here we observe that at low molecular weights beads or particles can form due to insufficient molecular chain entanglement during the fiber fabrication process. We and others have found that polymers with M_n < 10,000 form bead-like structures, while polymers with a relatively higher M_n (>50,000) form fibers during electrospinning at a low solution concentration [48]. All three PCL-based polyurethanes generated in this work demonstrated a number-average molecular weights of 50,000 and more, well within the expected electrospinning fiber forming range.

Solution concentration, applied electric field, and distance between the spinneret and collectors are also important factors which can be tuned to change nanofiber structure and were therefore optimized. All samples showed good processability at a concentration of 40 w/v%. As shown in Figure 1, smooth fibrous structures without beads were formed with average diameters of 417 ± 20, 511 ± 30, and 1329 ± 61 nm for PCL-6.8, PCL-9.2, and PCL-11.3, respectively. Interestingly, polymers with extremely low (<4.4%) or high (>21.3%) hard segment contents did not form consistent fiber structures. Electrospinning PCL-21.3 resulted in a bead-like structure rather than a fiber mesh (Figure S7). This is because polymer chains tend to aggregate as the hard segment amount increases. In addition to polymer composition, polymer architecture is also considered an important factor in electrospinning. Figure S8 shows the SEM images of the electrospun fibers from the four-branched PCL. Even though the molecular weight is similar, no fiber structures were formed.

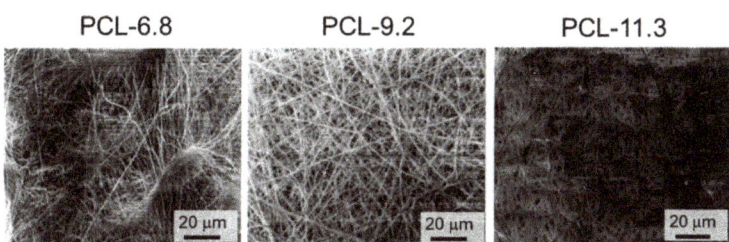

Figure 1. Scanning electron miscroscope (SEM) micrographs of fibers electrospun from PCL-6.8, 9.2, and 11.3 (10 kV accelerating voltage).

3.2. Thermal Properties

The thermal properties of the nanofibers were characterized by differential scanning calorimetry (DSC) (Figure 2). The peak attributed to hard segment was observed around 150 °C (Figure S9) [49]. Interestingly, the T_m of all three nanofiber meshes were slightly lower than that of the corresponding pure PCL nanofiber mesh suggesting disruption of polymer crystallization through hydrogen bonding between the hard segments [50]. In the case of nanofibers, this trend becomes more significant because the molecular orientation generated during the spinning process can lead to the alignment of polyurethane molecules along the fiber axis. The preferred orientation of the molecules leads to the packing and aggregation of hard segments into hard-segment microdomains. As a result, the differences in T_m between linear PCL and the PCL-based polyurethanes are accentuated. We also evaluated the crystallization of PCL-based polyurethanes by cooling scan of DSC (Figure S10). The exothermic peaks of PCL, PCL-6.8, PCL-9.2, and PCL-11.3 were found around 30 °C, 30 °C, 38 °C, and 36 °C, respectively. Although the crystallization temperature of polyurethanes was higher than that of pure PCL, heat of crystallization decreased as hard segment was introduced.

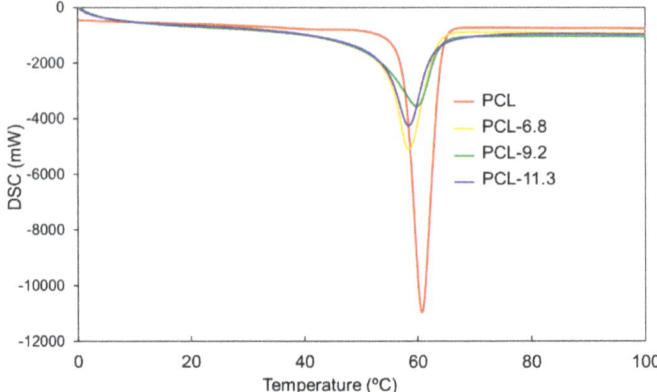

Figure 2. Differential scanning calorimetry (DSC) curves of electrospun fibers of pure PCL and PCL-6.8, 9.2, and 11.3.

To confirm the thermostability of the physically cross-linked polyurethane samples, samples were heated to near their T_m (around 60 °C) and their morphologies were analyzed by SEM. Figure 3 presents the SEM images before and after heating (magnified images was shown at Figure S11). Before heating, all the nanofibers showed the formation of uniform fibers. After heating, PCL-6.8 was melted, whereas PCL-9.2 and PCL-11.3 retained their fibrous morphologies. This result indicates that the physical cross-linking within PCL-6.8 was insufficient to maintain its morphology above its T_m. On the other hand, PCL-21.3 (>21.3%) broke after heating even though it had the highest hard segment content (Figure S12). These results suggest that careful selection of a soft:hard segment ratio is important for designing high-performance shape-memory nanofibers.

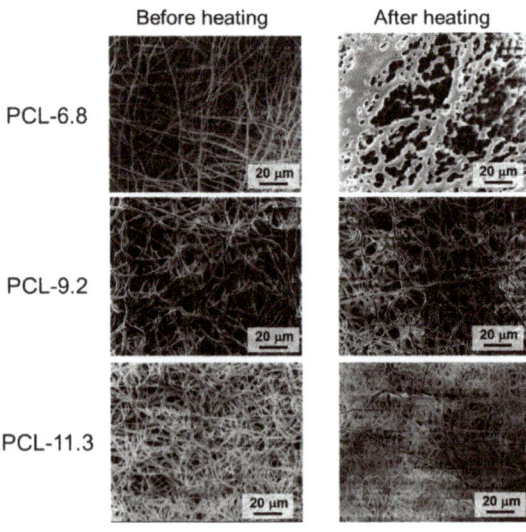

Figure 3. SEM micrographs of electrospun fibers of PCL-6.8, 9.2, and 11.3 before and after heating at 60 °C for 24 h.

3.3. Shape-Memory Properties

The shape memory capabilities of the nanofiber mesh were assessed by SEM and orientation analyses. Figure 4a shows digital camera, SEM, and cross-sectional images of PCL-9.2 nanofibers before deformation, after 300% deformation and fixation, and after shape recovery. The fibers showed different structures before and after deformation. The original, randomly oriented fibrous structure was easily deformed into a temporary stretched shape wherein fibers tended to orient along the strain direction. The oriented fibers recovered their original structure when the sample was reheated. Cross-sectional images of PCL-9.2 nanofibers during the shape memory cycle demonstrate that mean fiber diameter slightly decreased from around 600 nm to 500 nm after stretching (Figure 4b), and that the cross-sectional shape changed slightly from circular to ellipsoidal after stretching. Following shape recovery, the mean fiber diameter recovered to around 700 nm. The nanofiber structure was also stable after the shape memory test cycles at a microscopic level. Figure 4c presents a comparison between the R_f and R_r of the PCL-9.2 nanofiber mesh under different deformation rates (200, 300, and 400%). The R_f values were 88, 90, and 93%, and the R_r values were 100, 89, and 91% after 200, 300, and 400% deformation, respectively. In this study, we only showed PCL-9.2 data because PCL-6.8 melted above 60 °C, while PCL-11.3 is fragile and easily broken.

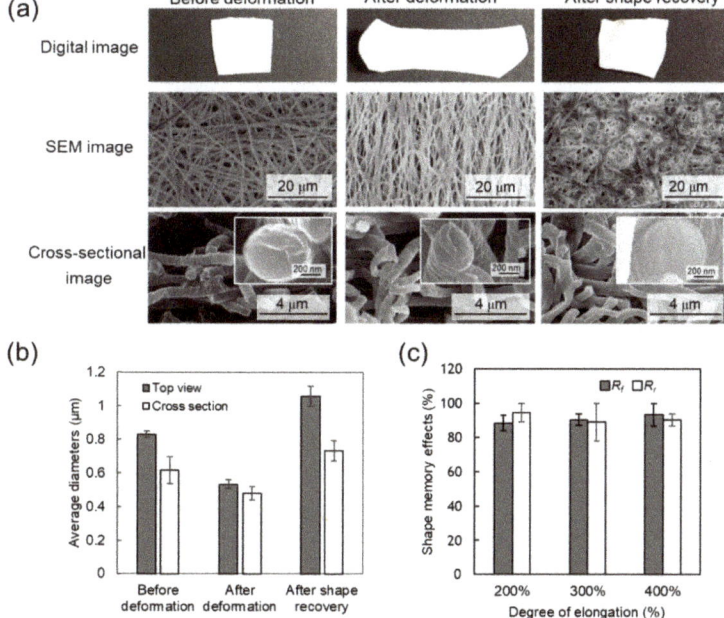

Figure 4. Evaluations of shape-memory effect of PCL-9.2 nanofibers. (**a**) Digital (top) and SEM (middle and bottom) images of PCL-9.2 nanofibers before deformation (left), after deformation (middle), and after shape memory recovery (right). (**b**) Average diameters of PCL-9.2 nanofibers calculated from the cross-sectional images before deformation, after deformation, and after shape memory recovery (n = 3). (**c**) Shape fixity rate (R_f) and recovery rate (R_r) of PCL-9.2 nanofibers under different elongation degrees (200%, 300%, and 400%) (n = 3).

To visualize the fiber orientation during the shape memory cycle, orientation analysis was performed using image analysis software (n = 3). The software evaluated the structure tensor of each Gaussian-shaped window by computing the continuous spatial derivatives in the x and y dimensions using Gaussian interpolation. For qualitative visual representation of the orientation, the grey-scale

SEM images were converted into color-coded images (Figure 5a). The fibers oriented along the deformation direction were assigned a ±90° orientation. The fiber orientation was clearly observed. For quantitative assessment, the average population of different orientation distributions of the fibers is presented in Figure 5b.

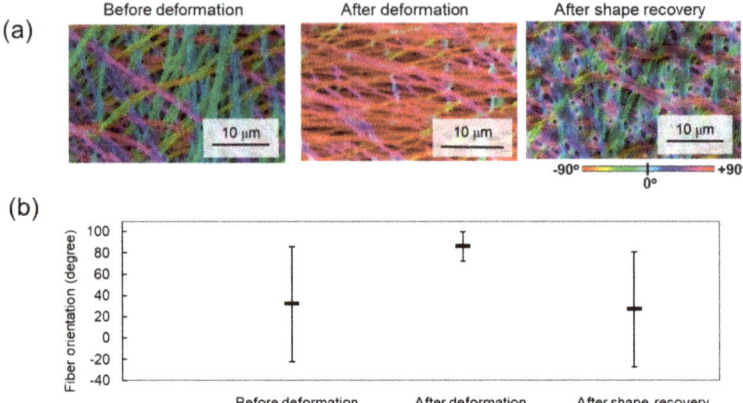

Figure 5. Orientation analysis of PCL-9.2 nanofibers before deformation, after deformation, and after shape memory recovery. (**a**) SEM images analyzed by Image J software for orientation evaluation and (**b**) fiber orientation distribution of nanofibers before deformation, after deformation, and after shape recovery.

3.4. Cellular Alignment

Matrix and scaffold topography are known to alter cell morphology and orientation. Recent studies have demonstrated that mesenchymal stem cells (MSCs) can sense the mechanical properties of their underlying culture substrate through integrin and focal adhesion kinase signaling to reorganize their actin cytoskeleton in response to extrinsic mechanical signals [18,48]. Therefore, we assessed the potential of our shape-memory nanofiber meshes to influence the MSC morphology. Human mesenchymal stem cells (hMSCs) were seeded onto the PCL-9.2 nanofiber mesh with and without deformation (300%). The cells on the random fiber mesh (non-stretched) spread extensively but with a random orientation (Figure 6), whereas the cells on the aligned fiber mesh (stretched) elongated along the fiber axis. The cells were seeded on the nanofiber before deformation and after shape recovery, respectively. Cellular alignments were then compared between these two conditions. Due to the long fibrous structure of the nanofibers, cells upon adhesion are geometrically restricted in the direction and orientation in which they can spread. Consequently, internal cellular structures like the cell cytoskeleton reflect the underlying material anisotropy resulting in cell alignment along the dominant fiber direction [51]. The cells cultured on the nanofibers were still alive after 2 days and the cell viabilities were similar to that on control substrate (glass substrate) (Figure S13). In this study, we cultured cells on the nanofiber before deformation or after deformation. Therefore, the effect of temperature was not considered. However, to observe how cells respond to dynamic change of the substrate during shape-memory activation is more attractive from the viewpoint of a mechanobiology study. Therefore, the next step of this study is to adjust the shape-switching temperature of fibers to 37 °C and examine dynamic cell culture on them. The shape-switching temperature of SMP can be possibly controlled by its molecular nanoarchitecture such as molecular weight, branched structure, hard/soft segment ratio etc [52].

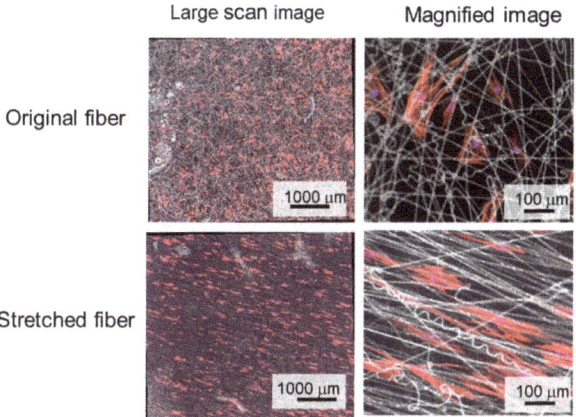

Figure 6. Adhesion morphology and alignment of human mesenchymal stem cells (hMSCs) on original (left) and stretched (right) PCL-9.2 nanofibers. hMSCs on each fiber type were stained with f-actin (red) and nucleus (blue). hMSCs were cultured at 37 °C for 24 h. Large scan (left) and magnified (right) images show global and local alignment of hMSCs.

4. Discussion

Thermally induced shape memory effect has the ability to recover their permanent shape from a temporary shape in response to biologically relevant temperature changes [13–15,17], thereby enabling the development of shape-memory cell culture substrates [13–15,17]. Nanopatterns can be programmed by mechanically embossing the desired topography into the polymer surface. Previous studies have demonstrated that cells cultured on these shape-memory surfaces can reorganize their cytoskeletons as well as alter their contractile direction in response to changes in material topography [39]. These studies have drawn much attention as they demonstrate a novel class of dynamic cell culture platforms to better understand the relationship between mechanical stresses and biological functions in a time-dependent manner. Through this work, it has been recognized that the mechanical interactions between cells and their extracellular matrix can regulate cell fate. In other words, cellular mechanisms are described whereby cells could sense, measure and respond to the space they were in, in particular to the rigidity of the surface the cells were on. By altering topographical cues in a controlled manner, these platforms enable researchers to recapitulate the dynamic nature of physiological conditions like wound healing, organogenesis and tumorigenesis in a dish. We also developed shape-memory microparticles by an in situ oil-in-water emulsion polymerization technique [16]. The particles with a disk-like temporal shape recovered their original spherical shape upon heating. Such particles can be potentially used for bioseparation, and in future, for drug delivery and immune engineering. In this study, shape-memory nanofiber meshes were prepared by electrospinning. Interestingly, polymers with extremely low or high soft:hard segment ratios were unable to form reproducible fibers as they demonstrated too few or too many inter-/intra-polymer interactions or entanglements. Smooth fibrous structures were successfully formed from PCL-6.8, 9.2, and 11.3, although PCL-6.8 melted after being heated above the T_m because the physical cross-linking within PCL-6.8 was insufficient to maintain its morphology.

To discuss the detailed mechanism of the shape-memory nanofibers, the microscopic shape-memory properties were considered. As hypothesized, the hard segments governed the mechanical properties of the permanent shape of the polymer network, and the soft segments as the reversible phase permitted shape memory properties. If the sample temperature exceeds the T_m, the crystalline region of the molecular chains will melt allowing one to be deformed. If the nanofiber meshes are stretched, the molecular chains will extend. Upon cooling to below the T_m, the molecular chains will crystallize. Resultantly, internal stress is stored in the extended fibers as the mesh is maintained in an intermediate semi-crystalline state. After reheating to above the T_m, the molecular chains become flexible and the nanofiber mesh recovers its original shape. Therefore, as hard segment content increased, shape fixity increased and shape recovery ratios decreased gradually. In this work, the fibers subjected to the deformation and recovery processes demonstrated microscopic changes in fiber structure as observed by SEM which matched macroscopic changes visible by eye. These results indicated that the shape-memory properties of the nanofibers were substantially determined by the internal structure of the PCL-based polyurethane but not by the macroscopic structural changes within the non-woven fabric. Compared to other shape-memory forms, many factors need to be considered to design shape-memory nanofibers. For example, fibrous materials have higher porosity, which can provide a low shape fixity ratio. It has also been proven that the electrospinning process produces a partial molecular orientation that can lead to an increase in the amount of the hard-segment phase, resulting in an increased recovery stress. However, the increase in the recovery stress adversely affected the fixing of the temporary shape, and thus, the shape fixity decreased. We also evaluated the recovery stress of PCL-9.2 after shape-memory activation. The value obtained from a tensile test was around 11 MPa (Figure S4). This value is relevant to previously reported values [42].

These unique and tunable aspects of shape-memory nanofibers may make them uniquely suitable for many biomedical applications. For example, their rapid recovery might be utilized in stents and endovascular thrombectomy devices in surgical approaches. In contrast, a slower recovery rate would provide an opportunity to study the long-term effect of dynamic matrix structure changes on cell behaviors such as cell differentiation and proliferation. Although here we only demonstrate cell culture on the shape-memory meshes before and after deformation, we believe that shape-memory meshes with transition temperatures within cytocompatible ranges may be utilized to control cell behavior through dynamic shape-memory changes.

5. Conclusions

We reported a novel strategy for the facile production of shape-memory nanofiber meshes. As demonstrated, smooth fibrous structures without beads were formed from PCL-based polyurethanes with different soft-to-hard segment ratios. The original nanofibrous structure easily deformed into a temporary shape, and recovered its original structure when the sample was reheated. A significantly high recovery rate (>89%) was obtained even when the mesh was deformed up to 400%. Furthermore, hMSCs aligned well along the fiber orientation when they were cultured on the meshes. Owing to their good biocompatibility, the proposed shape-memory nanofiber system would provide an opportunity to study the effect of dynamic matrix structure changes on cell behaviors. Moreover, another advantage of electrospinning is the possibility of encapsulating drugs in the fibers.

Supplementary Materials: The following are available online at http://www.mdpi.com/2079-6439/7/3/20/s1.

Author Contributions: Conceptualization, M.E.; Methodology, K.U.; Software, K.U., E.N., and K.T.; Formal Analysis, K.T.; Investigation, K.T.; Resources, M.E.; Data Curation, K.T. and E.N.; Writing-Original Draft Preparation, E.N.; Writing-Review and Editing, K.U., M.E., and A.K.; Supervision, M.E.; Project Administration, M.E.; Funding Acquisition, M.E.

Funding: This research was funded by JSPS KAKENHI (Grant Number 264086 and 26750152).

Acknowledgments: The authors thank the support from Namiki Foundry at National Institute for Materials Science (NIMS) for SEM measurements.

Conflicts of Interest: The authors declare no conflict of interest.

References

1. Gall, K.; Dunn, M.L.; Liu, Y.; Finch, D.; Lake, M.; Munshi, N.A. Shape memory polymer nanocomposites. *Acta Biomater.* **2002**, *50*, 5115–5126. [CrossRef]
2. Lendlein, A.; Langer, R. Biodegradable, elastic shape-memory polymers for potential biomedical applications. *Science* **2002**, *296*, 1673–1676. [CrossRef] [PubMed]
3. Nagahama, K.; Ueda, Y.; Ouchi, T.; Ohya, Y. Biodegradable shape-memory polymers exhibiting shpar thermal transitions and controlled drug release. *Biomacromolecules* **2009**, *10*, 1789–1794. [CrossRef] [PubMed]
4. Hu, J.; Zhu, Y.; Huang, H.; Lu, J. Recent advances in shape–memory polymers: Structure, mechanism, functionality, modeling and applications. *Prog. Mater. Sci.* **2012**, *37*, 1720–1763. [CrossRef]
5. Liu, Y.; Du, H.; Liu, L.; Leng, J. Shape memory polymers and their composites in aerospace applications: A review. *Smart Mater. Struct.* **2014**, *23*, 023001. [CrossRef]
6. Wang, J.; Quach, A.; Brasch, M.E.; Turner, C.E.; Henderson, J.H. On-command on/off switching of progenitor cell and cancer cell polarized motility and aligned morphology via a cytocompatible shape memory polymer scaffold. *Biomaterials* **2017**, *140*, 150–161. [CrossRef] [PubMed]
7. Tseng, L.F.; Mather, P.T.; Henderson, J.H. Shape-memory-actuated change in scaffold fiber alignment directs stem cell morphology. *Acta Biomater.* **2013**, *9*, 8790–8801. [CrossRef] [PubMed]
8. Bothe, M.; Emmerling, F.; Pretsch, T. Poly(ester urethane) with Varying Polyester Chain Length: Polymorphism and Shape-Memory Behavior. *Macromol. Chem. Phys.* **2013**, *214*, 2683–2693. [CrossRef]
9. Hsu, S.-H.; Hung, K.-C.; Lin, Y.-Y.; Su, C.-H.; Yeh, H.-Y.; Jeng, U.S.; Lu, C.-Y.; Dai, S.A.; Fu, W.-E.; Lin, J.-C. Water-based synthesis and processing of novel biodegradable elastomers for medical applications. *J. Mater. Chem. B* **2014**, *2*, 5083–5092. [CrossRef]
10. Chien, Y.C.; Chuang, W.T.; Jeng, U.S.; Hsu, S.H. Preparation, Characterization, and Mechanism for Biodegradable and Biocompatible Polyurethane Shape Memory Elastomers. *ACS Appl. Mater. Interfaces* **2017**, *9*, 5419–5429. [CrossRef] [PubMed]
11. Takahashi, T.; Hayashi, N.; Hayashi, S. Structure and Properties of Shape-Memory Polyurethane Block Copolymers. *J. Appl. Polym. Sci.* **1996**, *60*, 1061–1069. [CrossRef]
12. Leng, J.; Lan, X.; Liu, Y.; Du, S. Shape-memory polymers and their composites: Stimulus methods and applications. *Prog. Mater. Sci.* **2011**, *56*, 1077–1135. [CrossRef]
13. Ebara, M.; Uto, K.; Idota, N.; Hoffman, J.M.; Aoyagi, T. Shape-memory surface with dynamically tunable nano-geometry activated by body heat. *Adv. Mater.* **2012**, *24*, 273–278. [CrossRef] [PubMed]
14. Ebara, M.; Akimoto, M.; Uto, K.; Shiba, K.; Yoshikawa, G.; Aoyagi, T. Focus on the interlude between topographic transition and cell response on shape-memory surfaces. *Polymer* **2014**, *55*, 5961–5968. [CrossRef]
15. Ebara, M.; Uto, K.; Idota, N.; Hoffman, J.M.; Aoyagi, T. The taming of the cell: Shape-memory nanopatterns direct cell orientation. *Int. J. Nanomed.* **2014**, *9*, 117–126. [CrossRef] [PubMed]
16. Uto, K.; Ebara, M. Magnetic-Responsive Microparticles that Switch Shape at 37 °C. *Appl. Sci.* **2017**, *7*, 1203. [CrossRef]
17. Uto, K.; Mano, S.S.; Aoyagi, T.; Ebara, M. Substrate Fluidity Regulates Cell Adhesion and Morphology on Poly(ε-caprolactone)-Based Materials. *ACS Biomater. Sci. Eng.* **2016**, *2*, 446–453. [CrossRef]
18. Uto, K.; Aoyagi, T.; DeForest, C.A.; Hoffman, A.S.; Ebara, M. A Combinational Effect of "Bulk" and "Surface" Shape-Memory Transitions on the Regulation of Cell Alignment. *Adv. Healthc. Mater.* **2017**, *6*, 1601439. [CrossRef] [PubMed]
19. Naheed, S.; Zuber, M.; Barikani, M. Synthesis and thermo-mechanical investigation of macrodiol-based shape memory polyurethane elastomers. *Int. J. Mater. Res.* **2017**, *108*, 515–522. [CrossRef]
20. Cha, D.; Kim, H.Y.; Lee, K.H.; Jung, Y.C.; Cho, J.W.; Chun, B.C. Electrospun nonwovens of shape-memory polyurethane block copolymers. *J. Appl. Polym. Sci.* **2005**, *96*, 460–465. [CrossRef]
21. Kotek, R. Recent advances in polymer fibers. *Polym. Rev.* **2008**, *48*, 221–229. [CrossRef]
22. Zhuo, H.; Hu, J.; Chen, S. Electrospun polyurethane nanofibres having shape memory effect. *Mater. Lett.* **2008**, *62*, 2074–2076. [CrossRef]
23. Chung, S.E.; Park, C.H.; Yu, W.-R.; Kang, T.J. Thermoresponsive shape memory characteristics of polyurethane electrospun web. *J. Appl. Polym. Sci.* **2011**, *120*, 492–500. [CrossRef]
24. Okada, T.; Niiyama, E.; Uto, K.; Aoyagi, T.; Ebara, M. Inactivated Sendai virus (HVJ-E) immobilized electrospun nanofiber for cancer therapy. *Materials* **2016**, *9*, 12. [CrossRef] [PubMed]

25. Garrett, R.; Niiyama, E.; Kotsuchibashi, Y.; Uto, K.; Ebara, M. Biodegradable nanofiber for delivery of immunomodulating agent in the treatment of basal cell carcinoma. *Fibers* **2015**, *3*, 478–490. [CrossRef]
26. Brannmark, C.; Paul, A.; Ribeiro, D.; Magnusson, B.; Brolen, G.; Enejder, A.; Forslow, A. Increased adipogenesis of human adipose-derived stem cells on polycaprolactone fiber matrices. *PLoS ONE* **2014**, *9*, e113620. [CrossRef] [PubMed]
27. Hyysalo, A.; Ristola, M.; Joki, T.; Honkanen, M.; Vippola, M.; Narkilahti, S. Aligned Poly(epsilon-caprolactone) Nanofibers Guide the Orientation and Migration of Human Pluripotent Stem Cell-Derived Neurons, Astrocytes, and Oligodendrocyte Precursor Cells In Vitro. *Macromol. Biosci.* **2017**, *17*, 1600517. [CrossRef] [PubMed]
28. Stitzel, J.; Liu, J.; Lee, S.J.; Komura, M.; Berry, J.; Soker, S.; Lim, G.; Van Dyke, M.; Czerw, R.; Yoo, J.J.; et al. Controlled fabrication of a biological vascular substitute. *Biomaterials* **2006**, *27*, 1088–1094. [CrossRef] [PubMed]
29. Stankus, J.J.; Guan, J.; Fujimoto, K.; Wagner, W.R. Microintegrating smooth muscle cells into a biodegradable, elastomeric fiber matrix. *Biomaterials* **2006**, *27*, 735–744. [CrossRef] [PubMed]
30. Xu, C.; Inai, R.; Kotaki, M.; Ramakrishna, S. Electrospun nanofiber fabrication as synthetic extracelular matrix and its potential for vascular tissue engineering. *Tissue Eng.* **2004**, *10*, 1160–1168. [CrossRef] [PubMed]
31. Riboldi, S.A.; Sampaolesi, M.; Neuenschwander, P.; Cossu, G.; Mantero, S. Electrospun degradable polyesterurethane membranes: Potential scaffolds for skeletal muscle tissue engineering. *Biomaterials* **2005**, *26*, 4606–4615. [CrossRef] [PubMed]
32. Shields, K.J.; Beckman, M.J.; Bowlin, G.L.; Wayne, J.S. Mechanical properties and cellular proliferation of electrospun collagen type. *Tissue Eng.* **2004**, *10*, 1510–1517. [CrossRef] [PubMed]
33. Yang, F.; Murugan, R.; Wang, S.; Ramakrishna, S. Electrospinning of nano/micro scale poly(L-lactic acid) aligned fibers and their potential in neural tissue engineering. *Biomaterials* **2005**, *26*, 2603–2610. [CrossRef] [PubMed]
34. Jin, H. Human bone marrow stromal cell responses on electrospun silk fibroin mats. *Biomaterials* **2004**, *25*, 1039–1047. [CrossRef]
35. Li, W.J.; Tuli, R.; Huang, X.; Laquerriere, P.; Tuan, R.S. Multilineage differentiation of human mesenchymal stem cells in a three-dimensional nanofibrous scaffold. *Biomaterials* **2005**, *26*, 5158–5166. [CrossRef] [PubMed]
36. Dalby, M.J.; McCloy, D.; Robertson, M.; Agheli, H.; Sutherland, D.; Affrossman, S.; Oreffo, R.O. Osteoprogenitor response to semi-ordered and random nanotopographies. *Biomaterials* **2006**, *27*, 2980–2987. [CrossRef] [PubMed]
37. Silva, G.A.; Czeisler, C.; Niece, K.L.; Beniash, E.; Harrington, D.A.; Kessler, J.A.; Stupp, S.L. Selective Differetiation of Neural progenitor cells by high-epitope density nanofibers. *Science* **2004**, *303*, 1352–1355. [CrossRef] [PubMed]
38. Curtis, A.; Wilkinson, C. Nantotechniquies and approaches in biotechnology. *Trends Biotechnol.* **2001**, *19*, 97–101. [CrossRef]
39. Mengsteab, P.Y.; Uto, K.; Smith, A.S.; Frankel, S.; Fisher, E.; Nawas, Z.; Macadangdang, J.; Ebara, M.; Kim, D.H. Spatiotemporal control of cardiac anisotropy using dynamic nanotopographic cues. *Biomaterials* **2016**, *86*, 1–10. [CrossRef] [PubMed]
40. Matsumoto, H.; Ishiguro, T.; Konosu, Y.; Minagawa, M.; Tanioka, A.; Richau, K.; Kratz, K.; Lendlein, A. Shape-memory properties of electrospun non-woven fabrics prepared from degradable polyesterurethanes containing poly(ω-pentadecalactone) hard segments. *Eur. Polym. J.* **2012**, *48*, 1866–1874. [CrossRef]
41. Fejos, M.; Molnar, K.; Karger-Kocsis, J. Epoxy/Polycaprolactone Systems with Triple-Shape Memory Effect: Electrospun Nanoweb with and without Graphene versus Co-Continuous Morphology. *Materials* **2013**, *6*, 4489–4504. [CrossRef] [PubMed]
42. Ji, F.; Zhu, Y.; Hu, J.; Liu, Y.; Yeung, L.; Ye, G. Smart polymer fibers with shape memory effect. *Smart Mater Struct.* **2006**, *15*, 1547–1554. [CrossRef]
43. Barmouz, M.; Behravesh, A.H. Shape memory behaviors in cylindrical shell PLA/TPU-cellulose nanofiber bio-nanocomposites: Analytical and experimental assessment. *Compos. Part A* **2017**, *101*, 160–172. [CrossRef]
44. Kawaguchi, K.; Iijima, M.; Miyakawa, H.; Ohta, M.; Muguruma, T.; Endo, K.; Nakazawa, F.; Mizoguchi, I. Effects of chitosan fiber addition on the properties of polyurethane with thermo-responsive shape memory. *J. Biomed. Mater. Res. B Appl. Biomater.* **2017**, *105*, 1151–1156. [CrossRef] [PubMed]

45. Aslan, S.; Kaplan, S. Thermomechanical and Shape Memory Performances of Thermo-sensitive Polyurethane Fibers. *Fibers Polym.* **2018**, *19*, 272–280. [CrossRef]
46. Karger-Kocsis, J.; Keki, S. Biodegradable polyester-based shape memory polymers: Concepts of (supra)molecular architecturing. *Express Polym. Lett.* **2014**, *8*, 397–412. [CrossRef]
47. Lim, D.I.; Park, H.S.; Park, J.H.; Knowles, J.C.; Gong, M.S. Application of high-strength biodegradable polyurethanes containing different ratios of biobased isomannide and poly (-caprolactone) diol. *J. Bioact. Compat. Polym.* **2013**, *28*, 274–288. [CrossRef] [PubMed]
48. Maeda, T.; Kim, Y.J.; Aoyagi, T.; Ebara, M. The design of temperature-responsive nanofiber meshes for cell storage applications. *Fibers* **2017**, *5*, 13. [CrossRef]
49. Zhu, Y.; Hu, J.; Yeung, L.-Y.; Liu, Y.; Ji, F.; Yeung, K.-W. Development of shape memory polyurethane fiber with complete shape recoverability. *Smart Mater. Struct.* **2006**, *15*, 1385–1394. [CrossRef]
50. Ping, P.; Wang, W.; Chen, X.; Jing, X. Poly(e-caprolactone) polyurethane and its shape-memory property. *Biomacromolecules* **2004**, *6*, 587–592. [CrossRef] [PubMed]
51. Guan, J.-L. Role of focal adhesion kinese integrin Signaling. *Int. J. Biochem. Cell. Biol.* **1997**, *29*, 1085–1096. [CrossRef]
52. Li, F.; Hou, J.; Zhu, W.; Zhang, X.; Xu, M.; Luo, X.; Ma, D.; Kim, B.K. Crystallinity and morphology of Segmented Polyurethanes with Different Soft-Segment Length. *J. Appl. Polym. Sci.* **1996**, *62*, 631–638. [CrossRef]

 © 2019 by the authors. Licensee MDPI, Basel, Switzerland. This article is an open access article distributed under the terms and conditions of the Creative Commons Attribution (CC BY) license (http://creativecommons.org/licenses/by/4.0/).

MDPI
St. Alban-Anlage 66
4052 Basel
Switzerland
Tel. +41 61 683 77 34
Fax +41 61 302 89 18
www.mdpi.com

Fibers Editorial Office
E-mail: fibers@mdpi.com
www.mdpi.com/journal/fibers

www.ingramcontent.com/pod-product-compliance
Lightning Source LLC
LaVergne TN
LVHW072001080526
838202LV00064B/6814